THE IDEA OF CHINA

THE EAST ASIAN INSTITUTE OF COLUMBIA UNIVERSITY

The East Asian Institute of Columbia University was established in 1949 to prepare graduate students for careers dealing with East Asia, and to aid research and publication on East Asia during the modern period. The faculty of the Institute are grateful to the Ford Foundation and the Rockefeller Foundation for their financial assistance.

The Studies of the East Asian Institute were inaugurated in 1962 to bring to a wider public the results of significant new research on modern and contemporary East Asia.

THE
IDEA OF CHINA

MYTH AND THEORY IN GEOGRAPHIC THOUGHT

ANDREW L. MARCH

DAVID & CHARLES

NEWTON ABBOT   LONDON   VANCOUVER

0 7153 6507 X

Set in 11 on 13pt Baskerville and printed in
Great Britain by Latimer Trend & Company
Ltd Plymouth for David & Charles (Holdings)
Limited South Devon House Newton Abbot
Devon

Published in Canada by Douglas David &
Charles Limited 3645 McKechnie Drive
West Vancouver BC

CONTENTS

5

I

INTRODUCTION: GEOGRAPHIC THOUGHT IN THE CHINESE GREAT TRADITION

In the south-east quadrant of the Eurasian landmass, separated by deserts, mountains and seas from other old civilisations, lies China, yellow on the map. It is one of the five countries in the world with over 3 million square miles of territory and has by a wide margin the greatest population. It is high, dry and empty in the west, low, humid and crowded in the east, frigid in the north, torrid in the south; its big rivers let down into the Eastern and Southern Seas, and its broad plains often suffer flood and drought. So what?

This kind of geographic description, which can be indefinitely expanded, has various overlapping uses: for entertainment; for tourist or military intelligence; for social theory and myth. This book is concerned with the last sort: some of the ways in which geography has been used, theoretically and mythically, to answer the question 'What is China like?', ie, what does it feel like to be Chinese, and what will China do in the future?

By theory I mean simply a generalisation from a particular selection of facts, plus a causal interpretation. Most if not all of the social-geographic theories dealt with here have also a strong aspect of myth—that is, they contain a dramatic plot, a story or a pattern for stories, imbuing theory with value and meaning in reference to an implicit protagonist 'we' group. 'Myth' is not meant to be derogatory; a myth may be not false but obsolete, muddled, faithful to only some of the facts, damaging to other we-groups. In particular, the ones criticised here are against the interests of all but a small fraction of the world's population.

As for geography, although some geographers are concerned with physical science, and others do not distinguish their field from geometry and so study the formal properties of spaces whose content need not be specified, we are interested in geography here only as a social science, or rather as certain parts of a unitary inclusive '*super-science sociale*'.[1] Two central concepts are region and natural environment. A region is an area within which certain generalisations apply and a unit of geographic comparison. Natural environment is the physical universe with which societies interact but excluding the parts under their purposeful control—the tame or gardened parts. It is within patterns of physical and cultural regions that geographers and others have often set about explaining differences among societies by differences in natural environment, whether by a crude 'environmental determinism' or in a more sophisticated 'cultural ecology'.[2]

The method has many attractions, besides linking social science with the more prestigious and (often) better-funded natural sciences. Environmental elements vary over the world in easily stated ways, escape human control, yet are clearly important to man (eg, climate is obviously connected with dress, crops, architecture and so must affect economy, government, and sex life . . .); and since environment is by definition external to society, there is the possibility of an elegant and parsimonious one-way causation that is relatively easy to think about. Above all, there seems to be no alternative strategy of research that can promise more in the way of unified explanations of variation among societies.[3] Even if no single element such as climate, landforms or general location is enough, still the total mass of 'environmental' materials and processes constitute an impressive counterplayer for a given society and might well be expected to account adequately for much of its culture and development.[4]

Thus C. P. FitzGerald's *China* begins, as if it were a truism, with the observation that 'The history of every country is, to a great extent, determined by its geography'.[5] Seventy years ago Demolins wrote: 'If the history of mankind were to begin again, without any change in the surface of the globe, that history would in broad lines repeat itself.'[6] Nowadays we would be inclined to expand on the statistical view implied by 'in broad lines' to allow for the chance that actual history was a fluke, and thus say instead: if you set up the pieces and ran it all again not once but a thousand times, the outcomes would form regular and eventually predictable patterns—although any one trial, including what really did happen, would be indeterminate.

But although, in the mid-twentieth century, I find it inconceivable that world history would look random after a thousand runs, I do not accept that the explanation lies in the 'surface of the globe' or that it makes sense, in most of the inhabited earth, to divide a caused history from a causing globe in that manner. It is only in extreme negative situations that environmental causes—climate, landforms, soil, waters, biota, rocks—impose themselves as obviously overriding: that no great agrarian civilisation arose in Antarctica, or that fishing is not important in the Rub' al Khālī, requires no other kind of explanation. But in the parts of the world where the vast majority of the human race lives and where large, complex and progressive societies have developed, it is the presence of real cultural characteristics, not the absence of hypothetical ones, that needs accounting for, and here even all the environmental elements added together provide only a *necessary*, but never *sufficient*, cause. As Mao says, it is people, not things, that are decisive (cf chapter 5). Moreover, whatever the possibilities for an anthropologist with the far more numerous smaller and simpler societies about which data are available, conclusions about the role of environment in shaping past and present Chinese civilisation which emerge from comparisons with other cultures must be unsure or excessively abstract simply because there are too few cases enough like China; we would need the replays of history, or else other globes. Imaginary experiments, such as wondering what if north China had fifty inches of precipitation instead of twenty inches, if China were only the size of France, or were an archipelago, are indispensable to social-geographic thought but cannot make up for the lack of real comparative cases.[7]

More important, it is a sterile logic that orders the world in a two-term, one-way pattern of environmental cause→social effect; the price of such elegance is too high. Although probably no one today would subscribe to an uncompromising environmentalism, the strict Man/Nature separation underlying it is still very much alive, as will be seen in the sequel. Much more congenial to today's sense of reality is to conceive of the social situations before us as limited systems, nets, of more or less plastic, mutually responsive short-term and long-term events, whether 'natural' or human, some close to and some remote from our immediate interest. Man is not the one soft thing in a hard universe; the oak tree is not the cause of the squirrel. Within such systems, whether scholars or not we select a mix of elements on which to focus our attention, believing them to be strategic in the situation as a whole: ie, if they change, with them change also certain other elements in ways we wish to produce or explain. In most social situations only a few elements in the strategic mix belong to the 'natural environment' and it is gratuitous and misleading to single this out as a special category.

But the most serious objection to theory that finds the causes of history in the surface of the globe is that it is irrelevant or paralysing to a responsible social science whose primary concern should be actual people and their concrete existing problems. What is happening in the world now is change and revolution, and natural environment changes only very slowly in comparison with what matters in society—politics, war, personalities, economies, language, social organisation.[8] Writers to whom natural environment is strategic tend to de-emphasise social change or see it as superficial, and feed the same sort of complacent upper-class fatalism as racial determinism does. Social science needs to be oriented toward a future which, however it turns out, must be different from the past. With due attention to geographic and other necessary causes, the main stress must be on the experimental, dynamic and creative aspects of social situations.

In what follows, I have tried to pick out the most important sets of ideas that interpret China (or the Asia of which it is part) by geographic facts, especially location, climate, terrain, and resources. The great majority of them, in my opinion, turn out to be fallacious as theory and obsolete as myth; and broadly speaking it is the ideas represented by Mao Tse-tung's thinking that are most satisfactory in both respects—that the role of geography is best understood as secondary to human dynamism and class struggle. In the 'West', the concept *China* has strong geographic overtones—great size, teeming plains, the Yellow River—that seem to assort easily with notions about government, social organisation, and national psychology. For example, many arguments say more or less that geography produces an excess of uniform Chinese who consequently care little about the lives of individuals. Weird rationalisations can rest on such a foundation.

'I'm not a lover of war,' [Herbert D.] Doan [President of Dow Chemical] said. 'But this [napalm] is a fantastically useful strategic weapon. You know the people on that side of the world don't care about human beings the way we do here and they'll risk those human wave assaults.' 'Napalm,' he said, 'is the most efficient weapon to repulse the Asian onslaught.'[9]

The scholarly legitimations that make such reasoning plausible in America today are crumbling fast and I hope this book will speed the process.

Social theories, then, are not just games or art, but tools for actions with real consequences, dominated by mythic purposes. It is simply not true that one can 'describe processes of change without a deep commitment to values other than those of an objectivity that seeks to examine a subject from all points of view, and with a balance and perspective that correspond as closely as possible to the reality that is perceived by means of a consideration of the available evidence', as C. E. Black claimed recently in a well-received study of modernisation[10]—the whole preceding analysis to which this passage refers does embody major commitments that by no means grow self-evidently out of passively accepted data. He has found the Germany of 1933 to be one of the earliest countries to reach the last phase of modernisation, 'integration of society', one therefore in which the 'degree of integration is such that the pressure for the common welfare predominates over interest groups'.[11] Such 'objectivity' is ludicrous—Germany in the year Hitler became Chancellor the most advanced form of human society, the road to the future.

Internal logic and fidelity to fact by themselves are insufficient standards in social theory. Who is the implicit 'we', the fully human subject? And what is being said, not about the past, but about the present and the future? What myth is served, and for whom? A critique of social thought is alive and responsible only as it keeps these questions in sight.

Geographic thinkers in traditional China were of course far better acquainted with the concrete details of their country's natural environment and social organisation than were most of the non-Chinese authors with whom this book is concerned. A look at some of their ideas is an excellent starting perspective on how fact, logic and myth will interplay in reading the geography of China—or in any such broad social subject.

Traditional Chinese geographic thought was part of the ideological rationalisation and support of a stratified society—one with differential access to basic resources (means of production) and with class exploitation.[12] Such a system is maintained not only by ideology but also by violence. As to the structure of the society, I accept the view that 'traditional China consisted of a universal state characterized by a common high culture and a multiplicity of particular social systems of diverse local culture',[13] these local systems to be identified not with the single peasant village but with a group of villages linked by a common tie to a periodic market.[14] The gentry, constituting a couple of per cent of the population, is the privileged élite and main land-holding class in whose interest and by whose agency the state exists. Their ideology is in general the 'great tradition' upholding the image of a straight, hierarchical, and orderly society—'Confucian' ideology in a loose sense, roughly equivalent to the universal state's 'common high culture'. This society, as I understand it, was not *despotic* so far as the population at large was concerned, but was *oppressive*; that is, the state itself did not maintain a heavy coercive order throughout the society, or actually administer the affairs of the people, but the local social systems were hierarchical and authoritarian, favouring males, the wealthy, the older, and those with ties to the state. Chinese society took this characteristic shape by the Sung dynasty, around AD 1000, though it had stratification and state organisation from Shang times or earlier (mid-second millennium BC).

Although China's is undoubtedly the most abundant premodern geographic literature in the world in terms of both what was written and what is extant, it has been relatively little described in recent times.[15] In its narrowest sense, much of it is reference material and not intended for consecutive reading or as a source of ideas. Like most branches of literature, geography was written by members of the gentry—the literate élite—for readers like themselves, and often it was intended as an aid to administration, since the gentry typically aspired to or held government posts and needed to be familiar with the territorial administrative system and with the population, transport, economy, and so forth, of particular areas where they served. Similarly, readers of history—which consisted largely of the biographies and the administrative activities of these same people—required the same kind of information about past periods; but it was normally assumed that historical study too would find an ultimate justification in its application to contemporary society, and in a culture which had occupied the same core terrain since prehistory it was easy to feel that historical geography was essential knowledge for an administrator. A large part of administrative and historical geography was written under government auspices;[16] but other geographic writings merge in purpose and style with belles-lettres and are intended to improve the radius and differentiation of the sensitive reader's experience. All in all, the traditional geographic corpus contains a huge amount of material on a vast range of subjects: architecture, waterways, history, biography, ethnography, local economies, religion, education, geology, and so on. On the other hand, statements of interest from the viewpoint of geographic thought are often to be found in books of philosophy, history, or other subjects.

Modern cosmopolitan geography, as a social science in 'western' style, aims to be comparative. For the most part (in so far as it has any reality as a discrete field) it begins or ends as regional geography. General ideas are abstracted, in principle, from balanced and objective observations about many parts of the world taken together. In practice, as will be seen in later chapters by the example of China, it has often instead served its own myths of imperial and colonial involvement with the rest of the world. Since Chinese 'geography' was above all concerned with the single culture area of China, with little pretence at making balanced comparisons with other parts of the world at least until the nineteenth century, it is unlike modern geography. Virtually all knowledge about society was historical knowledge of China, as was argued by the eighteenth-century historian Chang Hsüeh-ch'eng and others before him.[17]

The geographic reality that the Chinese works refer to shows much more variety than western thinkers (as we shall see especially in the next two chapters) typically allowed it. Westerners considered China, and all Asia, as relatively uniform by contrast with the geography of Greece and the rest of the European-Mediterranean culture area. While it is true that China had less interaction with other great traditions than did the lands of the Mediterranean basin, still in terms of the folk cultures (eg, language, religion, lineage organisation, cuisine, variety of occupation) carried by the local social systems within China and around its margins, and in terms of overall geographical spread and variegation in climate, vegetation, landforms, and resources, it is hard to argue that China was not diverse.

Several salient points of Confucian geography's central myth are well summed up in this remark by Ch'iu Chün in the fifteenth century: 'As heaven and earth beget them, men have one same heaven but each a different earth [place, *ti*].'[18] 'Heaven' and 'earth' carry significance beyond the literal observation that the sky appears much the same to everyone while places on earth are so various. People are alike in that they are all supposed to share a natural propensity to form hierarchic family relations on the principle of subordination by generation, age, and sex, and to be receptive to carrying over the attitudes so learned into an acquiescence in or endorsement of the stratified society organised as a state. Dominance-subordination relations were not to be left to accidents of personality but were institutionalised so as to uphold the economic-political differential. This is the prime Confucian basis for the philosophical unity of man, the 'heaven' that all men share, and the essential part of human nature. It is a development in the 'common high culture' of the peasant kin, age, and sex ideology probably found in most of China's diverse local subcultures. Various often-repeated formulas sum it up, for instance *wu-ch'ang*, the Five Constants, the five basic asymmetrical pairs of social relations: father/son, prince/subject, husband/wife, elder brother/younger brother, friend/ friend, the first being most important; or *san-kang*, the Three Bonds, the first three of these five.

This is the first symbolic meaning of the 'one same heaven' that all humans share, in so far as they are fully human. The second is the top part of the social hierarchy—the emperor (the One Man, Son of Heaven) and those speaking in his name. 'Heaven is lofty, earth is low, thus *ch'ien* and *k'un* are settled. With low and high so set out, the noble and the base are positioned.'[19]

Earth, though, is various, and beyond the philosophical identity of their true potential best selves, so are people. 'People have no fixed substance—they change according to place [*ti*, earth].'[20] Again there is a significant plural meaning: *ti* is both the whole earth in its role of being subordinate to heaven, and parts of the earth, ie particular places, expressive of geographical differences in the conditions of human life. The diversity of places is to be subordinated to the unity of the sky.

Places are different from each other for natural reasons and for social reasons. The two are summed up together in the term *feng-su*: *feng*, 'wind, atmosphere', and *su*, 'custom, popular, vulgar'—ie, mores, local manners, culture, temperament. China's diversity is expressed in a proverb: 'A thousand miles and a different *feng*, a hundred miles and a different *su*.' The key explicative passage is in the geography section of the *Han-shu* (*History of the [Former] Han Dynasty*).

> All people have in their nature the Five Constants, but they differ in respect to hardness and softness, slowness and celerity, and the sounds they make; this is tied to the wind and breath of water and earth, hence it is called *feng*. And they are variable in what they like, dislike, take, and reject, and in their motion and rest; in this they follow their prince's emotions and desires, hence it is called *su*.[21]

Generally, then, mores are of two kinds: the *feng* kind are ahistorical, being part of the ever-present natural patterns of a place; and the *su* kind are historical and social.

A general term for the pattern of which *feng* (wind) is part also refers, in its primary sense, to vapours: *ti ch'i*, 'earth breath, atmosphere' or even 'local weather'.[22] It is climate as well as all the intangible qualities of a place.

> Broad valleys and great streams are variously formed, and the people living there have different usages [*su*]. (Note: ie, in what they like and dislike). In respect to hard and soft, light and heavy, slow and quick, they are differently tempered; [they like] different blends of the five flavours; their implements and weapons are differently fashioned; different clothing is suitable for them . . . The Chinese and the non-Chinese in the four quarters and the centre all have their own characters which cannot be made to change (note: this is due to the earth breath [*ti ch'i*]).[23]

The other aspect of mores (*su* in the *Han-shu* passage) refers to the differentiation of tastes and behaviour that comes from the influence of the upper social strata ('lord' or 'prince') independently of the natural character of a place.

> It is told that the king of Wu loved fine swordsmen, and his people were covered with scars; the king of Ch'u loved slender waists, and many in his palace died of hunger. There is a saying in Ch'ang-an: if high chignons are favored in the city, they will be a foot higher everywhere else; if broad eyebrows are favored in the city, they will cover half the forehead everywhere else; if big sleeves are favored in the city, they will take a whole bolt of silk cloth everywhere else.[24]

Natural geography (*feng* in this sense) does not set people on the right course, nor will local social influence necessarily do so; the effects of both may be trivial, morally indifferent, or downright evil. Virtue is introduced by a particular kind of social influence, one which addresses itself to people's heaven side rather than their earth side and leads them to recognise correct hierarchical authority and to exercise the proper forms and ceremonies for this purpose. Leadership of this kind is provided by sages—Confucius and the ancient, semi-legendary sage-kings—and their surrogates, China's 'superior men', 'scholars', true kings, virtuous officials. 'Wherever the superior man passes is transformed; wherever he abides is spiritualized. [His influence] flows together with heaven and earth above and below.'[25]

'Wind', as we saw, can stand for the natural character of places, but in other contexts it means the influence of these eminent persons. In the *Book of Changes* the hexagram *kuan* is interpreted as 'The wind blows over the earth . . . Thus the kings of old visited the regions of the world, contemplated the people, and gave them instruction.'[26] 'You are wind, and the people below are grass.'[27] Wind comes from afar, a personal instruction from (and about) 'heaven', while *ch'i* (in this sense) rises from the local earth.

According to the theory, with its blend of social and natural geography, sagelike leaders can arise only in one part of the world: its middle, China, the place where the environment is richest and most finely balanced, and the most accessible place from all the four quarters.

> In all the world, wherever the light of sun and moon reach, the Chinese occupy the middle ground. There the breath [*ch'i*] that acts upon living things is full and straight. The people have a well tempered nature and are intelligent; the land has rich and various products;—accordingly it begets sages and wise men, who continually propagate laws and doctrines, remedying the defects specific to the times, finding the usefulness of various objects . . . The precedence of sovereign and subject, elder and younger is established; the doctrines of the Five Constants and the Ten Duties [to the spirits, between sovereign and subject, etc] are perfected . . . China, in far-off ancient times, was in many respects like the barbarians of today. There are those who make their dwellings in nests and caves; there are those who bury without raising a gravemound; there are those who ball their food together with their hands . . . Their lands are marginal, the *ch'i* impeded; no sages or wise men are produced, and there is no one to reform the old ways; they are places where imperial instructions and admonitions are not accepted, and the rites and duties do not reach.[28]

This sense of social space—value concentrated in the Chinese middle and petering out in the four cardinal directions toward the lands of darkness and savagery—is schematised in the classics as a nest of concentric squares.

Thus the established view of China's and the world's geography is a mythical expression of the dominant Confucian ideology. It sees a tension between two opposing tendencies in society: on the one hand a constant drift toward degeneration and differentiation of social practices and morality under the pull of accidental local variations in the character of places and happenstance amoral leadership—a pull toward anarchism; on the other, a continually renewed conscious amelioration and unification led from the centre by sages and their representatives, in accord with the essential and universal nature of man which can flower only in a stratified state society. In its spatial expression it is the positive, moral, ordered centre striving to extend itself to all space, as against the neutral, orderless, amoral and not fully human edges tending to encroach upon and take over the centre. The centre should come materially to include all space—hence the exotic tribute articles regularly collected there; and all space should be made morally like the centre—thus the outward flow of personal admonitions, symbolised particularly by imperial tours of ceremonial inspection.

Was the intention of the myth to make the whole world alike and subservient? There were certainly tendencies in this direction. The uniformising measures taken by the First Emperor of Ch'in, who accomplished the first unification of China in 221 BC, have been regarded by Confucians and westerners alike as the very example of tyrannical despotism. He standardised laws, taxes, administration, coinage, weights and measures, the axle lengths of waggons, and writing, and carried out a purge of books and doctrines considered subversive. Though he acted violently and made little effort to pretend he was following the Way and Virtue for the improvement of his people, still much of what he did remained part of the Chinese governmental system.[29] A grim picture can emerge if one puts primary emphasis on these aspects of state Confucianism, especially as they bore on the gentry, and ignores the local social systems so that the individual seems to confront the state directly. For example, Étienne Balazs:

> If by totalitarianism is meant total control by the state and its executives, the officials, then it can indeed be said that Chinese society was to a high degree totalitarian . . . No private undertaking nor any aspect of public life could escape official regulation. In the first place there was a whole series of state monopolies . . . But the tentacles of the state Moloch, the omnipotence of the bureaucracy, extended far beyond that . . . This welfare state superintended, to the minutest detail, every step its subjects took from the cradle to the grave.[30]

The strength of feeling evident in such passages comes from a belief in their modern relevance: one can see 'the omnipresence, to this day [1957], of the old bureaucratic spirit in China',[31] and indeed the whole world is moving in the same direction, along 'the road to bureaucratic, technocratic state control'.

> In every country, whether underdeveloped or superindustrialized, it is always organized state capitalism that triumphs. This makes all societies alike, because it gives rise to the same tendencies. The group takes precedence over the individual, and the supreme power of the state is uncontested. Organization wins over competition, and everyone prefers to give up his freedom and submit.[32]

To cite one further example, the Chinese nightmare of a less serious writer such as C. Northcote Parkinson rests on a like perception. What he calls 'that "Great Similarity" which we have every reason to dread'[33] refers to a Chinese phrase, *ta t'ung*, that has retained much of its vigour from traditional into modern times. It occurs in the *Book of Documents* where Karlgren translated it as 'great concord', and more importantly in a late section of the *Book of Rites* with Taoist and anarchist overtones.[34] K'ang Yu-wei (1858–1927) used *ta t'ung* for his utopia, a world society that would be quite homogeneous: the harmful divisions of mankind by nation, class, race, family, and so forth, would cease to exist, all would use one language and one system of weights and measures, and an elected parliament would govern.[35] *Ta t'ung* is set as a goal in Sun Yat-sen's words to 'San min chu i', the national anthem of republican China, and it is translated 'world peace'. Mao Tse-tung, too, uses *ta t'ung* for the future 'realm' or 'world' that mankind is someday to enter; the official translation calls it Great Harmony and a note says: 'It refers to a society based on public ownership, free from class exploitation and oppression—a lofty ideal long cherished by the Chinese people.'[36] With this diversity of political context, the phrase carries no fatality to be dreaded, and it can just as well mean a co-operative and just socialism as a repressive and narrow Fascist statism, or even anarchy.

In practice there were strong ideologies and organisations countering the ruling myth more or less effectively in this huge country—popular as well as high-culture Taoism and Buddhism, rebellious secret societies, close-knit clans, separatist political leaders along the borders. Even the dominant Confucian policy was not to meddle in the everyday lives of the people, and in fact there was little possibility that the local official, lowest representative of the imperial government, could closely regulate the affairs of the couple of hundred thousand people in his district. The traditional *society* certainly had its oppressive features, but that is quite a different thing from pervasive oppression of the population at large by a central bureaucratic 'state Moloch'.

The Chinese were not unique in thus spatialising a myth expressive of the principal values of their dominant culture. 'Even the most intelligent groups,' writes Erikson, 'must orient themselves and one another in relatively simple subverbal, magic design. Every person and every group has a limited inventory of historically determined spatial-temporal concepts, which determine the world image, the evil and ideal prototypes, and the unconscious life plan. These concepts dominate a nation's strivings and can lead to high distinction; but they also narrow a people's imagination and thus invite disaster.'[37] The Chinese theory was generally logical and broadly consonant with fact (more so than western theories about China, on the whole): as one travels from the Chinese core in any direction, environments are generally poorer and support less dense populations and less elaborate cultures and political systems. But there was propagandistic wishful thinking in the stress on the universal attractiveness of Chinese culture with its stratified state organisation and the unique virtue of its leaders. There is a striking contrast with the western thinkers we are about to consider in that the latter, guided by their own myths, usually regarded China as marginal, isolated, and uniform, far from ideal in its situation, climate or resources—an inferior setting for human society.

Notes to this chapter are on pages 122–4

2

THE MYTH OF ASIA

By the thirteenth and especially the seventeenth and later centuries when real information became available to Europeans, China had already long been categorised geographically as a sub-region of Asia sharing with the other sub-regions a common lot of 'Asian' characteristics. The strongest heritage of pre-modern western geographic thought with respect to China is simply this: China is in Asia. The old function of Asia has to a large degree merged with the modern idea of Communism in an indispensable counterpiece to the idea of Europe (western civilisation). And it is China that is most extremely Asian and un-Euro-American, at the opposite ends of the earth. This is the tradition that is appealed to by formulations such as 'Asian Communism with its headquarters in Peking China'. It is still used to help explain much of what is asked, expected, feared, and advocated with respect to China. But the region 'Asia' is a relic of history, and there is no cultural or historical entity that can rationally be subsumed under this single term.

A formal region is an area with spatial coherence—common location—and one or more other defining common characteristics.[1] Asia has one common characteristic besides spatial coherence: it is a continental landmass distinct from the surrounding oceans. But it is almost never spoken of in this primary sense. The ambiguity of the bounding continental shelves and islands, the equivocal place of Russia, and the impossibility of consistent separation from Europe, Africa, Australia and the

23

Middle East, mean that in practice definition is abandoned and 'Asia' becomes a shorthand substitute for a list of contiguous lands from Afghanistan to Japan south of the Soviet boundary and south-east to include Indonesia and the Philippines.

What is predicated about a formal region usually by implication holds for its parts as well (if Asia is hot, then so is India) except where there is aggregation (Asia may have a huge population but it does not follow that Laos does). It is natural but illogical to assume that characteristics other than the defining ones must be common to a formal region. There is even less reason to expect that what is true of one part must be true of the others, unless it is included in the original definition of the region. Yet a thoughtful geographer can say of the formal region that 'whatever is stated about one part of it is true of any other part'.[2] If an author can make this slip writing about regions in the abstract in a book devoted to the subject, it is easy to see how illogical concrete thought may be about so vague a thing as Asia. Asia, in fact, is more a literary and psychological construct than a geographical one, and by a kind of double synecdoche there is associative leakage from one part to the other by way of the whole; it is a composite assembled from Persia, Egypt, India, etc, and each of these countries is expected to share more or less in all 'Asian' traits, not just those originally derived from itself. Regionalisations like this one are also examples of Erikson's 'relatively simple subverbal, magic design', providing some of the most important frameworks for group-identification, the division of the human race into 'we's' and 'they's', and so, with all their connotations, they connect with the deepest social emotions and values.

From the Greeks onwards Europeans have always thought of the world's lands as composed of two or more continental parts.[3] A threefold division—Asia, Lybia (Africa), Europe—prevailed both in antiquity and in the Middle Ages. But the Greeks had at first preferred a division into only Europe and

24

Asia, counting Africa with Asia;[4] and both Augustine and Orosius, who accept the threefold division, mention a secondary tradition that divides the world in two, putting Africa, however, with Europe.[5] Thus the two basic regions have always been Europe and Asia, with Africa included now in the one, now in the other, but usually making a separate third region. In the Middle Ages this ecumene was commonly depicted as a T in an O (Fig 1), corresponding to Augustine's explanation:

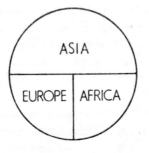

I say Asia meaning not that part [ie, Asia Minor] which is a province of this greater Asia but what is called Asia as a whole, which some count as one of two parts but most as one of three parts of the whole world [*totius orbis*] so that altogether there are Asia, Europe, and Africa: which they do not make by an equal division. For the part which is called Asia extends from the south through the east to the north; Europe, from the north to the west; and Africa thence from the west to the south. Whence two parts are seen to occupy half the world [*orbem dimidium*], Europe and Africa, whereas the other half, Asia alone. But the reason the former are made into two parts is that between them some of the Ocean's waters wash in, making our great [Mediterranean] sea. —Therefore if you divide the world into two parts, east and west [*Orientis et Occidentis*], Asia will be in one and Europe and Africa in the other.[6]

These continents had originally been nothing more than the various coasts of the Aegean and the eastern Mediterranean, but as geographical knowledge widened the names were applied to growing hinterlands. Asia's land boundaries were usually taken to be the Nile or the eastern or western boundary of Egypt, and the Don (Tanais) until 1833 when the Ural mountains and the Ural River came into use.[7] Enough writers counted Egypt with Asia to make the Nile, the pharaohs and so forth part of the traditional complex of 'Asian' traits.[8] From the end of the fourth century the three continents were more and more commonly assigned to the three sons of Noah, ancestors of all humanity after the Flood, so Asia was occupied by Semites, Africa by Hamites, and Europe by Japhethites; but the boundaries were fuzzy—Isidore gives Japheth half of Asia as well as all of Europe. Much later some took these as racial distinctions as well, the three continents being those of the yellow, black, and white peoples respectively.[9]

Overlapping and merging with the threefold continental division is the twofold one between Orient and Occident, East and West, with North as a rule being counted more western and South more eastern. Since these terms remain primarily directional rather than areal, no exact demarcation by a Don or a Nile is required. They escape clear definition as regions and are thus easier than the continents to detach from unequivocal reference to real places. As in the passage cited from Augustine, Orient or East is equal to Asia and Occident to Europe (now also covering America and the rest of the 'western' world) and Africa; but in the long accretion of values and associations to these words, Africa was not included with the West.

This classificatory rhythm by which the earth is divided into several formally equivalent parts is developed speculatively in the notion of undiscovered habitable continents or ecumenes in addition to Asia, Africa, and Europe. The idea goes back at least to Aristotle and persisted until the positions of the actual continents on the globe were finally ascertained. The south temperate zone of the eastern hemisphere was inhabited by the antoeci, and the north and south temperate zones of the western hemisphere by the antipodes and the antichthones respectively (though the use of the terms varied). Interest in these places was maintained in the Middle Ages despite risk of heresy, sacred history and church doctrine requiring that all mankind be descended from Adam and within reach of Christ's salvation.[10] The great discoveries verified the existence of new continents, and the same form of thought continues today in the awareness that there must be rational beings elsewhere in the universe than on earth, some or all cut off from us by chasms of space and time as untraversable as the torrid and frigid zones of the old geography. Thus the concept of remote peoples possibly more advanced than themselves has always been present to Europeans, and Europe (like the individual countries within Europe) has always been classified as one in a list of sibling regions, at best by its own efforts temporarily *primus inter pares*. Inseparable from the Europeans' comparative viewpoint has been the sense that their own achievements were without final validity, being always subject to overshadowing by known or unknown civilisations outside Europe. This constant relativisation, especially *vis-à-vis* the East where through most of history the real rivalry lay, produced a social space loaded with competitive instability, in strong contrast to the paternally centred Chinese world space.[11]

It was within this geographic-conceptual schema, under the major headings of 'Asia' and 'Orient, East', that the European idea of China took shape. Long before there was more than one or two sentences' worth of knowledge (even fabulous) about China itself, the genus into which new information would be fitted was ready prepared in the European mind.

The most important Asian themes are already set out by the Greeks, especially Herodotus, Hippocrates, and Aristotle. Asian rulers are too rich: the fabulously wealthy Croesus was king of Lydia and Midas of the golden touch was king of Phrygia, both in Asia Minor. Herodotus begins his *Persian Wars* by telling how the Greeks and the Asians abducted each other's women: the Phoenicians carried off Io, later the Greeks kidnapped Europa and Medea, and finally Priam's son Alexander took Helen. 'In what followed,' Herodotus goes on, 'the Persians consider that the Greeks were greatly to blame, since before any attack had been made on Europe, they led an army into Asia . . . The Asiatics, when the Greeks ran off with their women, never troubled themselves about the matter; but the Greeks, for the sake of a single Lacedaemonian girl, collected a vast armament, invaded Asia, and destroyed the kingdom of Priam.'[12] Thus Asians, in contrast to the Greeks, do not care about individuals, or for individual feelings and relations. Xerxes, having constructed a bridge over the Hellespont to bring his enormous army from Asia into Europe, consulted the Greek Demaratus on what resistance the Greeks were likely to offer against his invasion. Demaratus said that the Greeks would resist being enslaved even if they could muster only 1,000 men to oppose him. Xerxes laughed. 'How could . . . even 50,000, particularly if they were all alike free, and not under one lord, . . . stand against an army like mine? . . . If, indeed, like our troops, they had a single master, their fear of him might make them courageous beyond their natural bent, or they might be urged by lashes against an enemy which far outnumbered them.'[13] Asia has huge multitudes of slavish people. Herodotus' Asia was above all Persia; but he too considered Egypt to be part of Asia, and his long account of this land, with its pyramids, canals, wisdom, and antiquity ('there is no country that possesses so many wonders'), the source of much that was Greek, added greatly to the idea of Asia.[14] India, too, appears in Herodotus: it is a land of multitudes of people ('more numerous than any other nation with which we are acquainted'), and much gold, peculiar tribes and practices, and occupying the easternmost part of the ecumene.[15] Arabia contributes its spices and prodigies,[16] and to the north the Scythians and their neighbours introduce the theme of barbarous nomadic hordes,[17] precursors of the Huns, the Mongols, and the Turks.

The work *Airs Waters Places* associated with the name of Hippocrates treats at some length of the characteristics of

Asia especially in contrast with Europe, as the two continents 'differ in every respect'. 'Everything in Asia grows to far greater beauty and size'; Asia is 'less wild' and its inhabitants 'milder and more gentle'. 'The cause of this is the temperate climate, because it lies to the east midway between the risings of the sun, and farther away than is Europe from the cold. Growth and freedom from wildness are most fostered when nothing is forcibly predominant, but equality in every respect prevails. Asia, however, is not everywhere uniform.' It is fruitful and has good waters and forests; but 'courage, endurance, industry and high spirit could not arise in such conditions either among the natives or among immigrants, but pleasure must be supreme'.[18] The differences between Asians and Europeans Hippocrates accounts for by a combination of environment and institutions:

With regard to the lack of spirit and of courage among the inhabitants, the chief reason why Asiatics are less warlike and more gentle in character than Europeans is the uniformity of the seasons, which show no violent changes either towards heat or towards cold, but are equable. For there occur no mental shocks nor violent physical change, which are more likely to steel the temper and impart to it a fierce passion than is a monotonous sameness. For it is changes of all things that rouse the temper of man and prevent its stagnation. For these reasons, I think, Asiatics are feeble. Their institutions are a contributory cause, the greater part of Asia being governed by kings. Now where men are not their own masters and independent, but are ruled by despots, they are not keen on military efficiency but on not appearing warlike . . . Subjects are likely to be forced to undergo military service, fatigue and death, in order to benefit their masters . . .[19]

Europeans are also more courageous than Asiatics. For uniformity engenders slackness, while variation fosters endurance in both body and soul; rest and slackness are food for cowardice, endurance and exertion for bravery . . . Where there are kings, there must be the greatest cowards. For men's souls are enslaved, and refuse to run risks readily and recklessly to increase the power of somebody else. But independent people, taking risks on their own behalf and not on behalf of others, are willing and eager to go into danger, for they themselves enjoy the prize of victory.[20]

A rich land but a monotonous climate and soft, cowardly, slavish, pleasure-seeking people: although Hippocrates was thinking especially of the people of Asia Minor, this is recognisably the same Asia that would eventually extend all the way to the Pacific Ocean.

Similarly, Aristotle writes in an often-quoted passage:

> Those who live in a cold climate and in Europe are full of spirit, but wanting in intelligence and skill; and therefore they retain comparative freedom, but have no political organisation, and are incapable of ruling over others. Whereas the natives of Asia are intelligent and inventive, but they are wanting in spirit, and therefore they are always in a state of subjection and slavery. But the Hellenic race, which is situated between them, is likewise intermediate in character, being high-spirited and also intelligent. Hence it continues free, and is the best-governed of any nation, and, if it could be formed into one state, would be able to rule the world.[21]

Roman writers repeated similar themes—the Orient is soft and slavish, kings are revered like gods[22]—but for our purposes added nothing new of importance. All these themes, mingled with Biblical and other Christian matters, became prominent again in Europe from the late Middle Ages on and are quite direct sources of modern ideas.[23]

In Christian writings 'Orient' (East) continued ancient connotations of the rising of the sun and stars and was also coloured by the eastern focus of the Bible. The image of the Orient as old, wise, and spiritual shows already in Herodotus' Egypt and remains common today. It is equivocal in valuation since mature age may also be seen as decadence, wisdom as cunning, and spirituality as heresy. In Latin the symbolic associations are reinforced by the expression of East as literally rising, with forms of *orior*; the King James's Bible loses this reinforcement by the use of several terms where the Vulgate has this one verb, but the images of light, enlightenment, beginning are still clear enough: 'There came wise men from the east [*ab Oriente*] . . . we have seen his star in the east [*in Oriente*]' (Matthew 2.1-2); 'Arise, shine; for thy light is come, and the glory of the Lord is risen [*orta est*] upon thee' (Isaiah 60.1); 'Behold the man whose name is The Branch [*Oriens*]; and he shall grow up [*orietur*] out of his place, and he shall build the temple of the Lord' (Zechariah 6.12); 'Through the tender mercy of our God; whereby the dayspring [*oriens*] from on high hath visited us' (Luke 1.78). By comparison with the East, where salvation has its source, the West is late and secondary: 'For as the lightning cometh out of the east [*ab oriente*], and shineth even unto the west [*usque in occidentem*]; so shall also the coming of the Son of man be' (Matthew 24.27). At worst, in this pair Orient/Occident, the Occident could be not just passive but positively dark, evil, heathen.[24] In Carolingian times, the two terms (also in the forms *ortus* and *occasus*) could be divorced from actual regions of the earth, as for example in Rabanus Maurus (*c*776–856): 'If by Orient [*Orientem*] is meant the kingdom of God, then indeed by Occident [*Occasum*] is meant hell, which is so far removed from the seat of the blessed as when Abraham says, "Between us and you a great chasm is established." '[25]

East has also historiographical meaning. Europeans have repeatedly envisaged a focus of history that moves from east to west in the direction of the sun and stars. Such a view could be an optimistic one, as in the elegant oxymoron of Orosius in a passage that also recalls the slavery/freedom topos of Herodotus:

> Babylon had just been overthrown by King Cyrus at the time when Rome was first freed from the domination of the Tarquin kings. So indeed at one and the same juncture of times the former fell and the latter rose; the former then first suffered the dominion of foreigners, the latter then first rejected the haughtiness of its own rulers; the former then as it were dying gave up an inheritance, the latter attaining manhood knew itself heir; at that time the Orient's empire set and the Occident's arose [*tunc Orientis occidit et ortum est Occidentis imperium*].[26]

Similarly, around 700 Pope Nicholas I says that the West is the new East since the coming of Peter and Paul: 'The Occident by their presence . . . became the Orient.'[27] Or the movement could be seen as decadence and a token of the approaching end of things; as in the Chronicle of Bishop Otto of Freising. Both knowledge and power, he says, have moved progressively west since the time of Babylon, and now, having reached Gaul and Spain, are coming to an end: 'And it is notable that all human power and knowledge begins from the Orient and has its end in the Occident, that the fickleness and evanescence of things may thereby be made manifest.'[28] The conception of a moving locus at which history is really happening at any given moment has remained strong in western thought. It has its most important modern developments through Hegel and Marx, and we shall encounter it again below.

The idea that the Orient was the source of history fits the usual placing of the Garden of Eden there. Although where the King James's Bible reads, 'And the Lord God planted a garden eastward in Eden' (Genesis 2.8) the Vulgate has 'Plantaverat autem Dominus Deus paradisum voluptatis a principio', most other versions, including the Septuagint, specify the east.[29] The tradition was continued into early modern times with the eastern setting of most utopias and imaginary lands.[30]

The East was also the home of the perversions of wisdom which are magic and heresy; Roger Bacon, for example, says the whole East is given over to and steeped in the magic arts, and Isidore puts the origin of magic in Persia. The image was reinforced by the East–West splits in the early medieval church.[31]

Outside of specifically religious contexts in the Middle Ages, Asia continued to be the great land of prodigy. India especially was the home of extraordinary peoples, accounts of whom dated back to classical times: the dog-headed, the one-eyed, the headless, and so on.[32] Besides the strange peoples and animals, and the spices, gold and silver, dyes, medicines, incenses, and gems that fill the Asian lands,[33] there are also the tribes of Gog and Magog, shut up behind the Caucasus by Alexander, who were sometimes identified with the Tatar invaders of the thirteenth century,[34] and Prester John, a wise and powerful Christian king thought to rule some part of Asia.[35]

32

In the twelfth century Gerald of Wales devoted several chapters of his *Topography of Ireland* to a nightmarish summary of the horrors and liabilities of the East: silks, treasures, fertility of soil and subtlety of mind are for him far outweighed by the threat of disease and death from ubiquitous poisons, from lethal animals, from the very air.[36]

This then is a good sample of the received ideas in the older European tradition about Asia and the Orient; this is what Europeans taught themselves to expect of the East. The elements form not a logical whole but a reservoir of not always compatible images that can be drawn on selectively to yield a favourable or—most commonly in recent centuries—an unfavourable interpretation of virtually anything that Asia may be found to contain. The leading themes can roughly be grouped under two headings. *Excess*: Asia is oldest in civilisation and religion, it is richest, biggest, most populous, has the biggest empires, the most astounding prodigies, the greatest rivers, the subtlest minds, the finest artificers, the worst dangers to the body and soul. All is overdone; the Greek mean is lacking. Other extra-European areas may have this or that extreme or prodigy, but only Asia has them in such overwhelming abundance. And *uniformity*: endless stretches of territory, monotonous climate, masses of unfree people, eons of finished history. Excess is dominant in the classical and medieval writings; uniformity, in which the idea of magnitude and a relative ignorance of geographic differentiations are reflected in the use of a single region-name, Asia, for everything to the east, gains importance later with the taking shape of the idea of Europe. In a way the two sets of ideas seem to contradict each other, since one aspect of excess is excessive variegation, seemingly the opposite of uniformity. But these are images, not primarily logical conceptions. And anything regarded as excessive in scale or even in variety is felt as deadening to human experience and sensitivity and thus is phenomenally monotonous.

It was only after medieval times that China began to contri-

c   33

bute significantly to this pool of images. Practically all ancient recorded knowledge about China was summarised in a couple of sentences by Yule:

> The region of the Seres is a vast and populous country, touching on the east the Ocean and the limits of the habitable world, and extending west nearly to Imaus and the confines of Bactria. The people are civilized men, of mild just and frugal temper, eschewing collisions with their neighbors, and even shy of close intercourse, but not averse to dispose of their own products, of which raw silk is the staple, but which include also silk stuffs, furs, and iron of remarkable quality.[37]

Nothing of importance was added for a thousand years. So when the world was divided into the continental regions and when this old division was assimilated into Christian thought, China, far from being as now the single most important component of the category Asia, scarcely existed in the European mind. Asia and the East were composites of elements drawn mostly from Egypt (and Ethiopia), the Middle East, Asia Minor, the Russian steppe, and India. What is known or imagined about nearer places can influence what is assumed about farther ones in the same direction, and prevailing traits are accentuated with distance; if the extreme north is ice and the extreme south is fire, the farthest east would naturally be most 'oriental'. Already in Herodotus we read that 'the extreme regions of the earth, which surround and shut up within themselves all other countries, produce the things which are the rarest, and which men reckon the most beautiful';[38] and Gerald of Wales remarks that places most remote from the centre are remarkable for their prodigies.[39] China, of course, lies at the extreme limit of the ecumene to the east.[40] In the two concentric circles of late-medieval geographical knowledge—the inner zone of direct acquaintance, the outer of second-hand literary knowledge—China takes its place in the outer one with such fabulous lands as that of the Hyperboreans to the north, the Mons Clima to the south, and Atlantis to the west.[41] Between Europe and China lay the whole of the rest of Asia, studded

34

with marvels and monsters, so that whatever was said of China could scarcely seem other than fabulous and dreamlike. The European mind was quite prepared for the antipodal Chinamen of recent centuries, the looking-glass people who do everything backwards, as it was prepared for the wild tales of Marco Polo and Sir John Mandeville at the end of the Middle Ages: China is at the eastern extremity of Asia.

What is the continuing use of this 'Asia', that it did not disappear along with other medieval conceits as the world became more fully known? As one follows the themes delineated above closer into modern times, the answer clearly emerges that the idea of Asia has been an indispensable element in the prevailing definition of Europe. Though substantiating details are endless, the pattern is simple: the idea of Europe, as of the West generally, has come into being as an adverse one, so to affirm a subject European 'we' (by contrast with a sub-regional subject *within* Europe) has meant opposing a non-European 'they' that in historical perspective is equated with Asia. Our modern 'Asia' is perpetuated not for science but on behalf of those strata whose care is to maintain the ideal of western civilisation and who benefit from its sacred myths of individualism, private property, and aggressive defence of liberty. To the centred world of China, civilisation was one and inclusive, distinguishable only from barbarism, and to be civilised was to participate in the Chinese high tradition. But for Europe (apart from the barbarism symbolised by Africa in the T-&-O maps, cf Fig 1) there were *two* ways of being civilised and the accepted definition of the European way required contradistinction from the Asian-Oriental-Eastern (and Communist) way. Thus, the concept 'Europe' has been more than anything else the negation of 'Asia'; Asia, *a fortiori*, is best defined as non-European civilisation. This myth is the backbone of European historiography and the source of much of its coherence and drama, and hence important to concepts of world history as well.

Though it has a history of its own, in prototype reaching back

35

at least to the *Persian Wars*, what mainly concerns us here is the myth itself, especially as it appears in authors of our own times. But they also weave earlier anticipatory hints into their own *post hoc* interpretations, seeing them as Europe's first awakenings to self-awareness.[42]

History began, accordingly, when the West took cognizance of itself against the backdrop of the East. Greece foreshadows Europe, and Greece differentiates itself from the East in the wars with Troy and the Persians and in Alexander's conquests. From Greece comes 'all that is most distinctive in Western as opposed to Oriental culture . . . it was with the Greeks that there first arose a distinct sense of the difference between European and Asiatic ideals . . . the European ideal of liberty was born in the fateful days of the Persian wars';[43] Greece waged 'against Asiatic imperialism the first fight for the freedom of Europe'.[44] The Greeks set the pattern of definition for the West. 'The Occident that the Greeks founded exists only to the degree that it directs its gaze toward the Orient, confronts it, comprehends it and rejects it, borrowing something that it reworks after appropriating and counterposes, illuminated, against the dark ground of Asia.'[45] The myth lends history an entertaining dramatic simplicity; thus, one author can write of Alexander's conquest as 'the necessary outcome' of the Persian wars, part of a work later carried on by Rome, and reaching ultimately into the nineteenth century when a 'last reaction of Greece on the Orient will give birth, in 1830, to the kingdom of the Hellenes'.[46] Greece in its interaction with the Orient did not itself remain pure: the rich nutriment of orientalism contained toxins as well as vitamins, and brought not only growth but corruption.[47] 'In the Hellenistic, then the Byzantine civilisations, the Oriental element in the end prevails over the Greek. The unity, the classical purity change and disappear. There is instead, not a decadence, but another composite therefore inferior state which in the final outcome will prove to be more Oriental than European.'[48]

Rome's continuation of the struggle with the Orient lay 'in the historic [*welthistorischen*] wars with the Carthaginians, the campaigns against the despots of the East and the duel between Augustus and Antony'.[49] When Augustus defeated Antony and Cleopatra at Actium, this was 'a battle of East and West, the final victory of the European ideals of order and liberty over oriental despotism', over 'the formless hosts of oriental barbarism'.[50] But victory was not, after all, final; Oriental forces crept back, and the bureaucracy of the late empire 'has its roots in the administrative traditions of the great oriental monarchies of Persia and Egypt, but if it was oriental in origin, it had been rationalised and systematised by the Western mind. Consequently, in spite of its faults—and they were many—it possessed something of the political spirit of Western civilisation'.[51]

Most of this is quite clearly retrojection on the part of modern historians searching for the defining antecedents of their Europe and western civilisation. From the Middle Ages on, though, the myth finds some contemporary footing. Europe or the West as a community came to consciousness at times when there was an Oriental challenge, variously the Eastern church and the Byzantine Empire, the Huns, and Islam.[52] To the images of Troy, Salamis, and Actium is added in this period the figure of Charles Martel at Poitiers (732), saving Europe from the followers of Mohammed (who was 'the answer of the East to the challenge of Alexander').[53] An eighth-century chronicle refers to Charles's forces as Europeans (*Europenses*)—the first and for long the sole time this word was used.[54] During the centuries of the Crusades, though the names of the continents had little force beyond their literal geographic reference, Orient and Occident (with Greece now included in the first) kept strong historical and emotional connotations.[55] The association of the sons of Noah with the continents took an important new turn in the twelfth century in a statement by Godfrey of Viterbo that seems to identify the descendants of Japheth, Europeans, with Christians.[56] The concept of Europe as the stronghold of Christianity took shape with the new Asiatic threats of the Mongols and especially the Turks. Aeneas Sylvius Piccolomini (Pope Pius II) wrote in 1453, hearing that Constantinople had fallen to the Turks, of how Christianity had been chased out of Asia and Africa, and was not left intact even in Europe; the Holy Land was shamefully lost, yet 'indeed it was more bearable to lose towns that we held among the enemy, than to be expelled from those cities that were founded upon our soil and previously belonged to the Christians'[57]—'our soil' is Christian Europe.

But with the end of the Eastern empire at the hands of the Turks, with the Renaissance interest in Greek and secular learning, and especially with the unprecedented quantities of new information about the world that were available to Europeans from the sixteenth and seventeenth centuries on, the perspective on the East shifted again. The Europe/Asia myth took on the historical and theoretical character it still has today, and instead of Asia America, first primitive then as competitor and possible heir, became for many thinkers the prime contemporary outside standard of comparison for Europe.[58] The image of this new Europe was imperial and industrial, enlightened and progressive; but the old Europe, the historical heritage of cultural values, was already 'made' by this time and its confrontation with the antique Asia lay in the past.[59] The idea of the West—Europe not in contrast to but together with America and also Canada, Australia, and New Zealand, ie, the areas of white European population—is heir to the conceptual slot once occupied by Europe, and the continuing vitality of the whole myth comes from the meaning it is given in its extension in the contemporary world. The old Asian modes live on in the modern 'Eastern World' and to an important degree are conceptually assimilated with Communism, the new Asian threat by which the 'West' (ie, those with a stake in maintaining the capitalist organisation of 'western' societies) invents itself anew and asserts its heritage in the hope of preventing revolution.

Communism is viewed with much the same stereotypes as the old Asia: it is civilised, but heretical in its materialism and its brain-washing social engineering; it is built on ruthless, enslaved, sub-human hordes. The old Africa role is inherited by the Third World. Russia, long equivocally poised between Europe and Asia, went east for good by turning Communist. 'The Bolshevik revolution seemed to announce in irrefutable manner that Russia belongs to a system of ideas and a conception of life not at all occidental but on the contrary genuinely oriental, much closer to the Asian ones than to the European.'[60] And beyond Russia more and more there looms China, always super-Asiatic and now super-Communist as well.

This interpretation of 'Asia' as being above all a negative construct whose purpose is to give shape to the idea of Europe is supported by the fact that Europe's most commonly stated defining qualities can be summed up under the two heads *moderation* and *diversity*—it cannot be accidental that these are the exact opposites of Asia's excess and uniformity. Moderation is exemplified in the passage from Aristotle already cited (p 30), where Greece is described as having the ideal intermediate position between Asia's heat and Europe's cold. The modern Europe is identified not with Aristotle's Europe but with his Greece. 'It is not at all forcing the meaning of the text,' writes Reynold about this passage, 'to see in it the expression of a kinship between the Greeks and the [European] barbarians. It is the notion of Europe that is in the process of gaining precision.' 'It gains precision,' he adds, 'against Asia.'[61] The theme is expressed not only in climate but more generally. Gerald of Wales writes of the West's 'golden mean in things, which supply our use in decent measure and suffice for the wants of nature', rejecting the poisonous opulence and the pomps of the Orient. 'There a superabundance of treasures; here a modest and honest sufficiency.'[62]

But diversity is much the stronger of the two themes in modern times, although it too has classical antecedents.[63] The idea that diversity creates value runs through practically all realms: in economics, multiplicity of products and occupations and inequality of income; in politics, multiplicity of parties and interest groups; in geography, variety of resources, terrains, climates; in the natural as well as the human world, evolution through war, struggle for survival; in history, variety of periods and foci. For Europe, a contrast with Asia (or later with the Communist world) is I think always stated or implied. So Friedrich von Schlegel, in a course of lectures in 1810-11, said:

> The consequences of the *Völkerwanderung* are immeasurable, for the whole of modern history; all that has developed in the last millennium and a half through the noble competition of so many and such great nations and forces has been realized solely because of it. If the *Völkerwanderung* had not happened, if the German peoples had not succeeded in ridding themselves of the Roman yoke, if rather what was still left of northern Europe had been incorporated into Rome, here too the freedom and individuality of the nations erased, and everything with the same homogeneity turned into provinces, then that splendid competition, that rich development of the human spirit in the newer nations would never have taken place. And it is precisely this richness, this diversity [*Mannigfaltigkeit*], that makes Europe what it is, that gives it the advantage of being the most advantageous site of man's life and culture. There would be no such free and rich Europe, but instead only One Rome in which everything would be melted and dissolved together, and instead of the rich European history, the annals of the single Roman Empire would provide us with a counterpart to the sorry uniformity of the Chinese chronicles...
>
> Asia, one might say, is the land of unity where everything is developed on a large scale and in the simplest relations; Europe is the land of freedom, that is, of cultivation through the competition of separate and multifariously individual forces.[64]

John Stuart Mill's essay *On Liberty* contains a classic statement of the argument, contrasting European diversity with a warning evocation of China's supposed *ta t'ung*, while at the same time justifying the continuing poverty of England's lower classes:

> The greater part of the world has, properly speaking, no history, because the despotism of Custom is complete. This is the case over the whole East ... We have a warning example in China—a nation of much talent, and in some respects, even wisdom ... remarkable, too, in

the excellence of their apparatus for impressing, as far as possible, the best wisdom they possess upon every mind in the community, and securing that those who have appropriated most of it shall occupy the posts of honour and power. Surely the people who did this have discovered the secret of human progressiveness, and must have kept themselves steadily at the head of the movement of the world. On the contrary, they have become stationary—have remained so for thousands of years; and if they are ever to be farther improved, it must be by foreigners. They have succeeded beyond all hope in what English philanthropists are so industriously working at—in making a people all alike . . . and these are the fruits . . .

What is it that has hitherto preserved Europe from this lot: What has made the European family of nations an improving, instead of a stationary portion of mankind? Not any superior excellence in them, which, when it exists, exists as the effect not as the cause; but their remarkable diversity of character and culture. Individuals, classes, nations, have been extremely unlike one another: they have struck out a great variety of paths, each leading to something valuable . . . Europe is, in my judgement, wholly indebted to this plurality of paths for its progressive and many-sided development.[65]

Europe owes its civilisation to the 'cross-fertilization' in Europe among 'Hebrew religious thought, Greek humanism and philosophy, and the Latin power of ordered drill and legal organization', according to Sir Ernest Barker, and he goes on to evaluate also the 'gifts' of the Celts, the Teutons, and the Slavs to the 'general inheritance of Europe'.[66] A geographer similarly stresses the diversity of influences that have converged in Europe and the variousness of the inter-communicating sub-environments in which they were nursed to fruition.[67] Examples of such interpretations could be multiplied indefinitely. The point here is not whether they are wrong or right—though let me suggest that they incline toward tautology in that a given 'contribution' is likely to be labelled both a cause of Europe's progress toward civilisation and a part or evidence of that civilisation. The point is that Europe's self-congratulatory diversity is made to stand out against Asian, and especially Chinese, uniformity. Gift, contribution, blend, stimulus, dialogue, synthesis: such are the terms used to represent European diversity; but in Asian contexts, mixing is usually stated in pejorative and subliminally racist expressions like hybrid or syncretic. Thus Dawson writes of the Roman empire, 'we might have guessed that this spiritual deficiency would lead to an infiltration of oriental influences', and contrasts Christianity with 'the cosmopolitan world of religious syncretism in which Greek philosophy mingled with the cults and traditions of the ancient East'.[68]

Diversity and moderation combine in a model Europe composed of elements each of which is moderate and whose total number and range of variation are moderate; thus while they can fruitfully interact no one of them can dominate, subordinate, or eliminate others and so diminish the diversity of the whole. Primitive and backward societies (old 'Africa' and the modern Third World) are inferior because lacking in diversity of cultural and environmental elements. Extra-European civilisation—'Asia'—is inferior because of the immoderate dominance of some element; the social and natural scales are too large, the texture is not fine enough for European-type man. From the Enlightenment on, 'European' writers seem to have in the back of their minds some such schema as this, in which 'diversity' is the key to value and progress, and may be absent either by 'deficiency' or by 'excess' (see Fig 2).

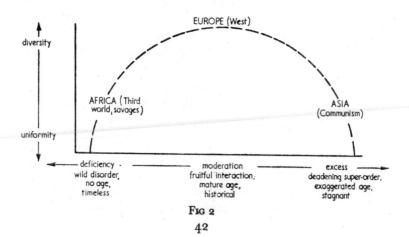

Fig 2

42

In Erikson's terms, the 'subverbal magic design' to which Europe orients itself is a moderate number of diverse, hierarchically equivalent, interacting compartments, in contrast to China's nest of concentric squares (see Fig 3). China's pattern

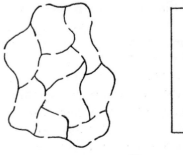 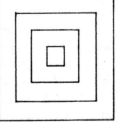

FIG 3

is centralised and mandala-like, its feeling of stability enhanced too by its squareness. Its historical dynamic is tension between centre and edges; the monuments and scenes of its whole history are all located together in the middle. The European pattern is comparative and premissed on rivalry; its historical dynamic is friction among the parts, struggle, and an ambiguously shifting focus. The design exists at many levels and in many domains, political, religious, and so forth; as a schematic map, it represents the various nations of Europe or the West, or the old systems of continents and ecumenes, or the nation-state system of the contemporary world, or the super-national blocs and groupings of states. The monuments and scenes of its history are scattered following the shift of focus; something like China's centredness would be obtained if for example the Pyramids, the Acropolis, Bethlehem, the Baths of Caracalla, the Arch of Triumph, and Valley Forge were all located within the territory of the thirteen original American states.

Just as in the case of China, this 'magic' design is plugged in to Europe's (the West's) ruling ideology. The pattern is a direct psychogramme of a world characterised, for its Euro-American subjects, by range of possibility and hence choice, enterprise, escape from determinism—in a word (one that sums up the whole group of values), freedom. It is also a world marked by relativism, impermanence, and competition, and where war and war preparations of whatever kind are always justified by reference to this same freedom; the great milestones in its mythic past are the anti-Oriental actions noticed above: Marathon, Salamis, and so on.[69]

The geographic doctrine most appropriate to this design is possibilism, a refinement of environmental determinism that holds that, 'Nature does not drive man along one particular road, but... offers a number of opportunities from among which man is free to select.'[70] The American possibilist Isaiah Bowman writes: 'Man has an exceptionally high power of selection—he may take or leave elements of the environment with a freedom not enjoyed by other forms of life.'[71] But in application it is only Europe and European man that enjoy this freedom: 'the natural environment', Bowman says, 'is always a different thing to different groups'.[72] 'Part of mankind is subject to "iron physical laws", but the part of mankind that is less directly affected by them is precisely the part that has developed the highest and most complex types of civilisation.'[73] Such civilisation is found only in parts of western Europe, the United States, Canada, Argentina, Australia, and New Zealand.[74] In other words, environmental determinism, the absence of freedom from 'extreme conditions' and 'iron physical laws', still rules the 'Africa' and 'Asia' ends of the world spectrum.

Much of the thought underlying the arguments to be taken up in the next chapters is summed up in this design and the values contained in it.

Such patterns and ideologies are not to be confused with the material reality and the concrete political arrangements behind the expressions 'Europe' and 'the West'; in this respect they hide more than they reveal. But for our purposes here, in order

44

to come to grips with European geographic thought concerning China, we must use the viewpoint of Curcio (and many others) that, 'Europe has been and still is above all an idea'[75]—that it makes sense as a real and definable region for other purposes need not concern us. But even if 'Europe' is more than an idea, Asia is not; it is simply a negative category without concrete basis, imposed from outside, for ulterior reasons, on a huge and heterogeneous area. A realistic regionalisation of the Eurasian landmass in terms of any definite and useful criteria would have to include not two but at least half a dozen hierarchically equivalent historical and cultural subdivisions, such as the areas we call South-west Asia, South Asia, South-east Asia, Central Asia, Siberia, East Asia—and Europe.

> Soul, consciousness, idea of Europe, has been said and is said . . . In the same manner a soul of the Orient is often spoken of, to indicate not so much and not solely a type of civilisation and conception of life, but especially that diffuse, buried and elusive feeling that gives to that world its tone, color and meaning, and that is not properly represented in any of its particular aspects but is to be understood in its own unitary ideal and spiritual interpretation . . .[76]

Such an Asia is nothing but a poorly analysed aspect of the intellectual history of Europe; and there is no other Asia.

Notes to this chapter are on pages 124–9

3

GEOGRAPHY, CHINESE
CIVILISATION, AND
WORLD HISTORY

In the eighteenth century and the early nineteenth a recognisably modern concept of China began to be precipitated out of the vague inherited 'Asia' as the foundations of the present social sciences were being laid down by such men as Montesquieu, Turgot and Hegel. In this chapter we shall see how these three used 'Asian'-style geography in their influential interpretations of China and will look also at subsequent efforts to solve the partly geographical problem of integrating China into a consistent world historiography. The westerners who were making the history in the societies that produced these ideas were new, industrial at home and imperial abroad and caught up in the struggles of revolution and reaction that still continue today. These three prominent theorists are selected from a number who helped reformulate the old Asian stereotypes of excess and uniformity, extending their geographic dimension, applying them more specifically to China and making them compatible with the waxing business conception of the West's place in the world.

MONTESQUIEU

To Montesquieu, China was and remained a sub-category of Asia, and the rather large amount of new concrete information

46

about the country that was by then available was assimilated deductively into his thought within this framework. The essential fact about any society was to him (as to many Enlightenment thinkers) its government, and governments are of three sorts: republican, monarchic and despotic. It is in Asia that 'despotism is, so to speak, completely at home [*naturalisé*]'.[1]

In a republic, the people as a whole or in part are sovereign; in a monarchy, one man rules but under fixed and established laws; in a despotism, one man rules not by law but by his own capricious will.[2] The principle that makes a republic work is virtue, ie, not moral virtue but devotion to country and to public interest. The monarchy runs on honour, individual thirst for rank and glory, while despotism operates by fear. This classification of governments is not logically exhaustive, but is intended to be empirical; the republican governments are those of Greece and the Roman republic, monarchies are the governments of most of the post-Roman states of Europe, and despotic states are prototypically those of Asia—but in fact, besides isolated other examples of despotism as in the cases of Henry VIII of England[3] and the Natchez Indians,[4] most peoples live under despotic government, and it is only in Europe that the rare accident of a 'moderate government'—a republic or a monarchy—occurs.[5] What we have, then, is good government in two variants, at home only in Europe, and evil Asian despotism into which the republic or the monarchy can degenerate if their principles become corrupt; but despotism is corrupt by nature and always tends to fall apart. The function of despotism in Montesquieu's thought is as concrete evidence that all government is not necessarily good but runs the risk of being evil even in Europe:

> The problem is not when the State passes from moderate government to moderate government, as from republic to monarchy, or from monarchy to republic; but when it falls and precipitates itself from moderate government to despotism.
> Most of the peoples of Europe are still governed by manners. But if

by a long abuse of power or by a great conquest despotism were to become established to a certain point, neither manners nor the moderating influence of climate would hold up, and in this fine part of the world human nature would suffer, at least for a time, the insults to which it is exposed in the three others.[6]

His concern was especially with the over-absolute government of the French monarchy, and the old connotations of 'Asia', plus the actual facts of government in a few cases, could supply him with an exemplification of despotism that was systematic and concrete, not accidental or occasional.

Most of the qualities of despotisms fall within our two categories of excess and uniformity. Despotisms are excessive by implication of their contrast with 'moderate' states; they and savage peoples alike are cruel and harsh in their punishments, while only in moderate governments, where since life is sweet the threat of death is an adequate deterrent, can there be gentleness and pity.[7] In despotisms when the people are mutinous they carry things to extremes, but in monarchies such excess is rare.[8] People in despotic states are fearful, ignorant, beaten down, and (by implication) animal-like; the goal of the government is tranquillity—not true peace, but a silence of cities about to be occupied by an enemy, ie, terror.[9] Unlike monarchies, despotisms do not allow diversity of laws in the various provinces but are uniform throughout;[10] and they lack diversity in time as well, as their manners and customs never change.[11]

Seeking to explain Asia's despotism, Montesquieu begins with climate, whose empire is 'the first of all empires', ie, social determinants.[12] Basically he accepts the old equation of east with south[13] and thinks of Asia as hot. People in hot climates, he argues from a fanciful physiology (contrasting them with those in cool climates), are weak, lazy, and as timid as old men; they are very sensitive to pleasure and pain, sensual, and given to passion and crime.[14] China is an example of a climate where desire is so strong that it is considered a marvel of virtue if one is left alone in private with a woman without raping her.[15] Perhaps this is why the climate in China favours immense fertility, though the reason is also the fact that the people depend largely on a diet of fish.[16] In these hot climates, where people are cowardly, despotism is the norm; in cold climates, people are brave and free.[17] But Asia has, Montesquieu realises elsewhere, cold climates too, and he quotes accounts of extreme cold in East Asia reaching south to the latitudes of southern France.

Some Asians, then, must be free? But 'I reason thus: Asia has no real temperate zone, and places situated in a very cold climate there immediately touch those which are in a very hot climate, that is to say Turkey, Persia, the Mogol [India], China, Korea and Japan.'[18] In Europe, on the other hand, the temperate zone is of very great extent, and the climate changes by imperceptible gradations from warm to cold. In Asia, because there is no temperate zone, strong and warlike nations are right next to effeminate, lazy and timid ones which they therefore conquer. In Europe, neighbouring countries have about equal courage. Here is 'the great reason for the weakness of Asia and the strength of Europe, for the liberty of Europe and the servitude of Asia'.[19] The Tartars, natural conquerors of Asia, exercise a despotic empire over the conquered peoples to the south, and this despotism is extended back again to the north so that they themselves are enslaved; and Chinese colonists bring north the spirit of the Chinese government, as do those of the Tartars who are sometimes driven out of China— they carry back to their deserts a 'spirit of servitude acquired in the climate of slavery'.[20] Thus, even though Asia is not actually all hot and able to produce despotism by pure physiology, it might as well be because it has no freedom anyway.

A secondary reason for Asian despotism lies in terrain. Asia has broader plains than Europe, and is divided into larger blocks by its seas; its rivers do not constitute such barriers as Europe's. Consequently large empires form in Asia which are necessarily despotic, because all large states are despotic: only fear can keep them from breaking up. Europe is compartmented by its terrain into a number of medium-sized states where government can be by laws, and this situation has produced a genius of liberty, but in Asia there reigns 'a spirit of servitude which has never left it; and, in all the histories of this country, it is not possible to find a single trait marking out a free soul: there will never be anything but the heroism of servitude'.[21]

The government of China has in practice many good features to mitigate its despotism. Thus though the climate induces lust, women are kept shut up so that their morals are actually admirable;[22] though the emperor is also high priest (powers of state and religion not being separated as in monarchy), he is considered father of the people[23] and 'there are books which are in everyone's hands to which he himself must conform. In vain did an emperor [the First Emperor of Ch'in?] attempt to abolish them, they triumphed over his tyranny'.[24] Moreover, lands made and maintained by human industry call for [*appellent à eux*] moderate government: the main three such are Egypt, Holland, and the Yangtze delta. The preservation of this enormous fertile part of China 'demanded rather the usages of a wise than those of a voluptuous people, rather the legitimate power of a monarch than the tyrannical power of a despot'; power had to be moderated (as in Egypt and Holland), since the administration of such lands could not be abandoned to caprice and neglect. 'Thus despite the climate of China, where one is naturally inclined [*porté*] to servile obedience, despite the horrors consequent upon the too great extent of an empire, the first legislators of China were obliged to make very good laws and the government was often obliged to follow them.'[25] Furthermore, since China's climate gives the women an unexampled fecundity, the population increases despite tyranny; and there being, as in all rice-growing countries, frequent famines, groups of bandits are generated who turn revolutionary, and thus the emperor is forever in danger of being overthrown and killed if he does not govern well. Such 'particular and perhaps unique circumstances have brought it about that the government of China is less corrupt than it should be'; but all of them are accidental and he insists that they cannot affect the nature of the government which is and remains despotic, ruling by fear.[26]

With his idea of 'laws' Montesquieu argues that society, like nature (though not to the same degree), is intelligible by cause-and-effect reasoning. But his explanations taken separately appear categorical, and he gives no rule for combining them into a consistent body of theory which would explain why now one cause, now another comes to the fore. 'A number of things govern men: climate, religion, laws, maxims of government, examples of past things, mores, manners . . . In each nation, in proportion as one of these causes acts with more force, the others yield to it by the same amount.'[27] Accordingly, general statements are frequently softened by terms equivalent to 'tend', 'inclined', 'other things being equal'. 'In mechanics,' he says, 'there are frictions which often change or halt the effects of theory: in politics [ie, social science] there are as well.'[28] Hence in sum Montesquieu is left with a loose indeterminism that allows him to adjust his explanations to whatever representation he wishes to make. China is an Asian despotism, no matter in how many respects it may seem not to conform to the theoretical model of a despotism, no matter what restraints may seem to cancel the operation of Asian social or environmental factors causing despotism. That the principle of despotism is fear 'does not prove . . . that in a particular despotic State people have fear, but that they should have: otherwise the government will be imperfect'.[29] What he is doing, within the context of the France of his day, is providing a reasoned argument for his own ideal of a state: the limited monarchy with 'freedom' protected by division of powers among king, bourgeoisie and hereditary nobility;[30] and with wealth assured by a vigorous imperialistic commerce with the non-European world of America, Asia and Africa.[31] The nobility (his own class) should not engage in commerce[32] but is still an essential part of a monarchy, which without it would be despotism. 'The nobility . . . enters in some manner into the essence of monarchy, whose fundamental maxim is: no monarch, no nobility; no nobility, no monarch, but one has a despot.'[33] This is probably why Montesquieu is so insistent that China be counted despotic: it has no nobility (worth mentioning), and he will not admit that nobility has no necessary function.

Thus though most of what he actually says in detail about China is on the other side, the country is and remains a despotism, to be understood by reference to general geographic causes of Asian despotism regardless of the concrete evidence of its own geography and government.

TURGOT

In various works dating from around 1750,[34] Turgot sketches a theory of human progress of which geographical ideas, although little elaborated, are an essential part. The theory was never worked out in full consequence and detail, but despite some uncertainties the argument is clear enough. His emphasis was different from Montesquieu's: he wanted to establish a model which would account for the progress that occurred, as he saw it, at certain times and places but not at others. Again Greece and Europe find their definition over against an 'Asia' invented for that purpose and which then comes to include China; and again the differences between the two are explained by reference to geography, though to Turgot terrain rather than climate seems to be decisive. Progress to him has the broadest sense, including especially progress in thought—Bacon, Descartes, Newton, and so forth—but also in art and literature, technology, commerce, and institutions; in general, increasing happiness.

Here is a piecing together of his model, or story. The first states were the product of conquest and hence despotic, and they arose in the Orient, Asia, where the too great extent of the plains and distance between natural obstacles to movement facilitated conquest. These empires were big and populous, and they formed so early that there had not yet developed laws or other institutions by which they might be governed with regularity and moderation—hence they were ruled in the easiest manner, by an arbitrary military despotism. Here—in, for example, India, Egypt, and China—were the first advances in science and art. But progress was soon arrested as languages became fixed in too primitive a stage, and science, under the protection of the state, became venerable, mysterious, monolithic and conservative; despotism thus fosters uniformity, lethargy and servility, and stifles progress. A society like China was simply too successful too soon in instituting peace, reason and justice, and therefore it was bound to stagnate.

But in Greece, in Renaissance Italy, and (by implication) in Europe generally, there was no such stagnant uniformity. Greece was a host of small and weak states, continually fighting, associating, trading and mixing, but incapable in early times of being formed into a single empire because the terrain was all broken up among islands, mountains, and seas. Here progress could be resumed.

> The only reason Greece so surpassed the Orientals in the sciences she had from them, was that she was not subjected to one single despotic authority . . . If [she had been so united and] the law-giver had been Pythagoras' disciple, the sciences of Greece would have been forever limited to a knowledge of this philosopher's dogmas which would have been erected as articles of faith. He would have been what the celebrated Confucius has been to China. Fortunately the situation in which Greece found herself, divided into an infinity of little republics, left to genius all the liberty, all the concourse of efforts [*concurrence d'efforts*, ie, competition and co-operation] that it requires.[35]

Turgot's view seems to have been that a similar situation prevailed in Europe until quite close to his own times, and anticipating Hegel, Marx and Spencer he accepts all the violence, evil, and injustice that progress requires. Such was the only possible way in which the human race could move toward its perfection and happiness. A military spirit fosters freedom; crime and passion—beyond the intention of their none the less guilty agents—engender progress. The very virtues of China make it possible to despise it: it attained too early to the paradise of reason, the *ta t'ung*;

> thus the passions multiplied ideas, extended knowledge, perfected minds in the absence of reason whose day was not come and which would have been less potent if it had reigned sooner.
>
> Reason, which is justice itself, would not have taken from anyone what was his, would have banished forever war and usurpations, would have left men divided into a host of nations separated one from another and speaking different languages.—Limited as a result in its ideas, incapable of any sort of progress of mind, science, art, or polity, which is born out of the meeting of geniuses gathered together from different provinces, the human race would have remained forever in mediocrity. Reason and justice better heeded would have fixed everything, as has more or less happened in China; but what is never perfect must never be entirely fixed. Passions, tumultuous and dangerous, became a principle of action and consequently of progress; anything that pulls men out of the situation they are in and sets varied sights before their eyes, extends their ideas, enlightens them, animates them, and in the long run leads them to the good and the true, where they are drawn by their natural inclination: like grain repeatedly shaken in a winnowing basket, that falls back by its own weight ever purer of the light straws spoiling it.[36]

There is a contradiction with Turgot's notions of the uniformity of great Asian empires; here he sees the formation of empires as progressive because it mingles diverse peoples. It is the lateness of the hearkening to reason and justice that is crucial—the later the better, and in Europe it is only recently that the violent passions of hatred and vengeance have begun to give way to virtues and gentler emotions.[37]

The pattern of Turgot's conception can be abbreviated in two diagrams, one the same as we saw in chapter 2 (p 43) and one its opposite; he has clarified their specific geographic meaning (see Fig 4). On the left distinct but communicating entities, and variety, diversity, freedom, competition/co-operation (*concurrence*), passion, war, and progress; on the right a single uniform entity, with peace, servility, etc, and stagnation.

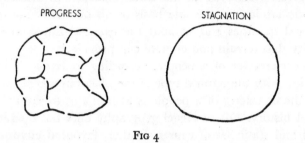

PROGRESS   STAGNATION

FIG 4

These images are perfectly congruent with Turgot's concept of the terrain of Asia on the one hand, and of Greece (by implication all Europe) on the other. It is evil passion, not reason or desire for progress, that forces the units in the left-hand diagram to interact. A variety of peoples in different places are driven violently together, but no one can dominate the other; such is the mechanism of human progress. China conforms to the stagnation pattern; since it is Asian, Turgot like Montesquieu includes it in his general explanation of Asian despotic empires and gives no consideration to the concrete facts of real Chinese geography or society. He too is saying far more about the existing class and national situation in Europe than he is about China, which remains essentially a negative projection.

HEGEL

Geography plays a minor but indispensable role in Hegel's philosophy of history. Idealist he may be in his concern with spirit (mind, *Geist*) as the primary historical reality, but at the same time he insists that spirit is always made concrete in a manner consonant with reason in real peoples at actual historical places and times. Geography is nature, and history is the progress of spirit away from nature through a series of societies in characteristic material environments laid out in order from east to west. Without geography, Hegel would be left assigning world-historical roles to the various peoples in arbitrary and chaotic fashion on the sole basis of his own estimate of their internal qualities and without the external fitness and plausibility that terrain and climate can provide.[38]

The character of a people, according to Hegel, is in close relation with the natural type of the places in which they live; and the character of a people is as much as to say its role in world history. But to Hegel geography does not operate as a rigid and static set of causes. Rather, favoured environments constitute a chain of increasing potentialities which lead into each other historically and logically and which are realised successively in the course of time by the world-historical peoples. These environments in which history has been possible are limited to Asia and Europe (including the Mediterranean coast of Africa). The torrid and frigid zones are excluded because men there are compelled to pay attention to nature at all times, and cannot detach themselves to build their own spiritual world apart from nature. And the south temperate zone is too split up, so only the north temperate zone in the old world has been a true theatre of history. The centre of world history, that without which it would be inconceivable, is the heart of the old world: the Mediterranean Sea, where three continents interacted.

Hegel distinguishes three essential geographic forms: arid uplands, the alluvial plains of great rivers, and seacoast. But these categories as it turns out are really applied *only to Asia*, and are evidently derived from Asia; America and Africa are excluded from history, and Europe's terrain (like its temperate climate) has a mild, in-between character without sharp contrasts of the three types, and especially without the opposition of upland and alluvial plain. Thus the effect of the geographic analysis in Hegel's system is to explain why he excludes most of the world from history, and more importantly why within the historical zone the Asian lands may be represented as inferior to Europe, whose geography is so subtle as to disappear in effect vis-à-vis the processes of history—a foreshadowing of the geographic possibilism we discussed above.

Asia then (omitting the frigid northern part) has the three geographic principles and their corresponding ways of life: the central Asian highland, with independent nomadic herders of patriarchal organisation; the plains of the great paired rivers—Yellow and Yangtze, Ganges and Indus, Tigris and Euphrates, Amu Darya and Syr Darya, and Kura and Aras—with agriculture, crafts, property, and state; and finally seacoast mixed with highland and plain in the Near East, ie Arabia, Syria and Asia Minor, characterised by trade, seafaring, and bourgeois liberties, and source of religions and political principles the best of which were passed on to be developed in Europe. Great Asian states like China, even if they lie on the coast, lack the outward call of the sea away from the limitation of the land—for them, the sea is nothing but where the land stops.[39] Thus China lacks the boldness, ingenuity, activity, and sense of adventurous freedom that Hegel associates with seafaring and sees in the Phoenicians and the Greeks.[40]

Except in the Near East, then, the true oriental elements are only the highlands and the great alluvial plains. But the highland peoples themselves are unhistorical, though they possess 'a powerful impulsion to alter their form' (I suppose by settling down and taking up agriculture in the river-valleys) and 'history takes its beginning from them'.[41] We are left with the alluvial plains as the only truly historical geographic element in Asia, and these Hegel describes incidentally to his discussion of Greece: Greece's islands, mountains, narrow plains, little valleys and streams, are all divided up yet all in communication by sea; the forms are various, without any single large-scale mass being predominant. 'We do not find that Oriental physical power, one stream such as the Ganges or the Indus in whose plains a homogeneous race never receives the invitation to change, because its horizon everywhere shows only the one same form . . .'[42] The combinations of terrain arrange themselves in a series of progressively finer texture, and (except for the last member, Europe) increasing complexity; Africa is wholly unhistorical, and only Europe is wholly historical. The series may be diagrammed as in Fig 5. The function of geo-

 AFRICA - highland

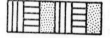 (EASTERN) ASIA – highland + alluvial plain

 NEAR EAST – same in finer texture + seacoast

EUROPE – all three elements so intimately mingled that they disappear as dominating forces leaving a possibilist world open to human freedom

Fig 5

graphy in Hegel's system, then, is to hold back all the areas of the world other than Europe from achieving full historicity.

During the course of world history, in Hegel's peculiar sense of the term, these areas are activated one after the other, only to be left behind as the spirit's spotlight shifts; in such a McLuhanesque linearity, there is never more than one protagonist country at a time. The East–West sequence explicitly follows the sun; Asia is the East not just relative to Europe but 'absolutely'; and it is in China, the easternmost part of Asia, that the great sunrise of human history occurred. China, the extreme East, is the 'Oriental principle' *par excellence*.[43] By the same token China is the most isolated and distant part of the historical world, and has not participated in the subsequent process of world history; the historical distance between China and Europe, at opposite ends of world history, is the greatest there is. Thus Hegel begins with China and passes on through India, Persia, Egypt, Greece (as usual, a major qualitative shift *vis-à-vis* Asia),[44] Rome, and Europe or rather German Christendom.

Hegel is not concerned with the development from arena to arena of techniques, science, administrative practice, or similar matters as such; his is not a brave progress toward human happiness through enlightenment such as Turgot saw in history. Rather he is interested in the relations between subject and substance—the individual self and the social and cultural matrix; subjective and objective forms of spirit. World history is a progress toward the full embodiment of objective reason in the state (ie, a politically unified national culture) and at the same time, for each member of society, toward a subjective sense of willing consent to the culture's norms which are wholly internalised and accepted as worthy of free allegiance. These two sides of the progress of spirit are represented (if we ignore the religion of the Mongols)[45] in geography by the principles of alluvial plain and highland respectively. Thus in Africa (highland) there is only wild subjective freedom; in Asia (highland plus alluvial plain) there is that, embodied in emperor and nomads, but also a massive substantiality from which the subject cannot detach himself—in China, the state, rational and moral but in which ethics are a matter for law and public administration and the citizens cannot have any conscience or subjective assent. The third principle—seacoast—has the function of generating motion (like passion in Turgot), especially by giving a progressive character to the bourgeois pursuit of profits, so that the other two principles can become fused to form, ultimately, the German Protestant state. The other 'world-historical' states in the East–West series fall in between, each with its own essential characteristic. This still fits Fig 2 (p 42 above), with wild subjectivity on the left and overwhelming substantiality on the right, and Europe in the middle, on top as usual.[46]

Hegel validates his system partly by its historical-geographical concreteness. He rigorously refuses to deal with fantasies, dreams (like the future world-historical role of America, or in fact anything in the future), and any kind of pre- or super-historical speculations. The forms of spirit, the stages in the inner and outer realisation of human freedom, must be, according to the rules of Hegel's game, concretely embodied in actual historical societies; they must really have happened, not remain mere abstract possibilities or conceivable tendencies. And since the unit of most intense human reality for Hegel is the state, to avoid abstraction he needs a whole historical state for each of his basic forms of spirit. All these stages of spirit, moreover, not only existed in the past but continue in the present as well; 'the phases [tendencies, elements: *Momente*] that Spirit seems to have left behind, are also still present in its depths'.[47]

There is no need nowadays to refute in detail Hegel's descriptive interpretation of China,[48] particularly here, since the connection with geography does not go beyond the general one set out above. 'Their main characteristic,' he writes in his concluding paragraph on the Chinese, 'is that everything belonging to spirit—free propriety, morality, feeling, inner religion, science and true art—is lacking.'[49] What he is writing about is not China but an aspect of the European mind, a mode of being of 'spirit' which seems possible to Europeans and which has taken shape round the European notions 'East, Orient, Asia, China'. Into these Hegel projected what he presumably felt and observed in his own world, that it is possible—even usual—for an individual mind to exist without the subjective sense of being a free self capable of moral choice, and that certain social institutions foster such a condition. Thus Hegel achieved the concreteness on which his system is predicated. But like Montesquieu's it is a spurious concreteness. Hegel's oriental pageant is imagery which has a certain literary force so long as one remains within the European context, but it is not, as it is intended to be, the faithful portrayal of real peoples. His moving spotlight of history shining on only one place at a time, coupled with his morose endorsement, in the name of progress, of selfish violence on the part of great men, reinforces a conception of humanity in which only bourgeois westerners are real (fully historical) and the rest of the world is subordinate and expendable.

WORLD HISTORY AND CHINA'S ISOLATION

Montesquieu and Turgot included China (and Asia generally) in treatments of world society which were presented as comparative although in fact they did not escape Europocentrism; but they did not try to establish spatio-temporal links, whether material or ideal, between China and the rest of world history. Hegel's system did not require that he show any material or concrete influence of China on the further progress of world history, but he did set up a linear development in the changing locus of the purely ideal World Spirit which at the beginning of history dwelt for a time in China before moving on west. Historians after Hegel were unwilling to follow him in this philosophical unconcern for concrete interconnections and so for the most part simply left China out of their 'world' histories.[50] Ranke omitted China together with all the 'oriental' lands because they were 'peoples of eternal stasis', and also not accessible to his method of painstaking critical use of sources; China, besides, had very few concrete connections with western history.[51]

This geographical isolation of China is an interpretive theme that recurs again and again in western writings of the last two centuries. 'A people in one corner of the earth,' Herder called the Chinese, 'placed by fate outside of the concourse of nations, and to that end fortified with mountains and deserts and a sea nearly without bays.'[52] And the geographer Richthofen saw 'the principal characteristic of the political and cultural history of the Chinese' in the 'seclusion in which this people has grown up', adding that only if one fully realised the contrast with Europe's situation could one appreciate what the Chinese had managed to achieve despite all.[53]

Isolation was imposed by difficult terrain and sheer distance, but the Chinese in their smug self-satisfaction made no effort to overcome it; on the contrary they further shut themselves off with the Great Wall, which 'represents China's heroic effort to perfect her almost complete natural isolation',[54] and when the Europeans came put off intercourse with them as best they could.

The argument that isolation explains China has two implications. First, that China has been handicapped by the absence of the specific and uniquely valuable elements ('contributions') making up western civilisation: Christianity, Roman law, Greek philosophy, etc. Second, more generally, that China has been without *any* comparable set of diverse, hence progress-generating elements and, deprived of the sort of fruitful interaction that Europe enjoyed, remained stagnant.[55]

Even at face value, this argument in either form actually says nothing at all about *China*, about the manners, economy, government, social structure or history of the Chinese; it only suggests that if China were located, say, in the Middle East it would be different. One might as well explain Europe by asserting that it would be more advanced if it were situated where the Indochinese peninsula is.

But the picture of China as cut off from the world is false from the start, and loaded with Europocentrism. Because of our indoctrination in them, the particular elements present in the 'West' seem natural and necessary and we persistently over-value them. In the matter of religion, for example, we assume that our heritage of Babylonian, Egyptian, Greek, Roman, Judaic, Christian, Druidic and whatever other strands presents a richer and more stimulating array than what was available to China; yet in China too, besides the native Taoism, Confucianism, and the old 'classical' religion,[56] there flourished at various times Judaism, Islam, Manichaeism, Nestorian Christianity, and the many diverse schools of Buddhism. Again, a Chinese scholar would be less likely than a European to know a 'foreign' language—but he had to learn classical Chinese just as the European cleric learned Latin, and whatever familiarity he had with Chinese 'dialects' acquainted him with languages differing from each other as much as do many of the Indo-European idioms of Europe. Similar arguments can be made respecting the economies, governments and other cultural elements that interacted in East Asia. In terms of sheer numbers, the fact that roughly one-quarter of the human race has, apparently for thousands of years, lived in what is now China makes absurd the connotations of smallness and sparseness in 'isolated' ('islanded', with a suggestion of *solus*, 'alone'). It is true that the Confucian tradition of state and gentry did minimise external contact in its written record, but ordinary people—traders, pilgrims, bandits, migrants—continually came and went.

In any event, since 1500, and especially in the last hundred years, there can be no dispute that a real world history in the sense of concrete, frequent, and significant interconnections among the great majority of human beings has existed and increasingly gained intensity. 'It is this new situation,' writes Barraclough, 'which makes the need for universal history—by which we mean a history that looks beyond Europe and the west to humanity in all lands and ages—a matter of immediate practical urgency'; without a rational grasp of universal history, we are not equipped to cope intelligently with 'some of the most important factors directly affecting our lives'.[57] But 'our lives' as far as I am concerned means the class interests of the majority of mankind, and new myths of world history must stand against the dominant 'western' bourgeois historiography, not just extend it.

There exist broadly three emphases to bringing Chinese (or any other) history into coherent world history. The first stresses such past contacts and interactions as did occur, in the Old World especially by way of the Central Asian nomads; the second holds that the present interdependence of the world imposes all the requisite unity on earlier history so that there is no need to magnify the importance of scanty earlier contacts; and the third focuses upon the essential similarity and solidarity of all human experience, regardless of contacts.

The peoples of Central Asia once aroused what has seemed to some a disproportionate amount of interest in comparison with the Chinese because they seemed to offer the possibility of concrete historical linkages among the various Eurasian civilisations, especially between China and the Mediterranean-European world.[58] Most notable was Frederick J. Teggart's effort to explain barbarian uprisings in Europe by correlating them with, among other things, events in China's western regions; and he asserts that 'if the history of Eurasia in general and of Europe in particular is to be understood, the history of China must be placed in the foreground'.[59] More broadly, Hodgson in advocating a Eurasian history as a legitimate simplification of the problem of world history would focus attention on themes directly or indirectly linking the various Eurasian civilisations, in practice again through the Central Asians; he suggests such topics as the Mongol empire, or the diffusion of Muslim mathematics to China and to Europe.[60] But such a world history would always remain thin and selective, and could not logically include more than a tiny part of the social, cultural, economic, and political history that would be chosen by almost any other criterion of significance. At most the insistence on concrete interaction for a world history including China is a useful antidote to Europocentrist and idealist historiography of Hegelian style.

The second approach is exemplified by Karl Jaspers. Interested in the historical parallels between China, India and the West, especially in what he calls the 'axial period' (*Achsenzeit*), the centuries around 500 BC when most of the great Eurasian religions arose, he finds evidence of causally significant interaction unconvincing and unnecessary for a conception of world history. 'They are three independent roots of a history that later —after interrupted separate contacts, conclusively only in the last few centuries, really only today—became a single unity.'[61] Practically all history is in fact written with some such built-in teleology, where items are included not for their contemporary significance or lateral interconnections alone but also with an eye to what the historian knows was coming. Who cares about Napoleon's childhood, except for what he did as a man? Similarly, American history is interested in the first Spanish settlement in Florida as well as the Massachusetts Bay Colony without feeling obliged to show concrete interconnection—later developments make the relevance. Exactly the same principle justifies the inclusion of China with Europe and the rest in a single world history. If it is objected that 'our' roots are after all in Greece, Rome, and Europe, there is no answer except that the 'we' that is thus defined is adverse, exclusive, and backward-looking, entirely inappropriate to the mid-twentieth century with its undeniable accelerating mutual dependence of all peoples.

The third possibility for an integrated world history stresses the ideal solidarity of the human species, or the concrete similarities in its experiences at all times and places regardless of past or present interactions among its parts. The unity of the species is expressed in various ways, but all urge that human existence as such contains its own meaning and justification to which respect is due. God made man in his own image, therefore the image of man is holy; ask not for whom the bell tolls; treat men as ends, never solely as means. Ranke writes, 'Because man dies, the individual life has value. All elements of the life

E   65

of nations must be regarded as independent developments, not [only] in so far as they serve a final development.'[62] Whatever their historical role, all human events have this aspect of being 'immediate to God'.[63] But this broad, vague compassion (in which China is implicitly embraced along with all mankind), though necessary to any acceptable historiography, is insufficient because unselective; it cannot decide what is important and then look for what caused it. A slightly different emphasis asserts the relevance to us of all human experience in virtue of the fact that our minds are capable of repeating the thoughts of other minds so that we enlarge our lives and learn our potentialities; the purpose of studying history is to pursue self-discovery through vicarious experience.[64] More generally, from a social science view of history any human experience (whether inside or outside the mind) yields potentially valuable data for making and checking explanations about societies, since all societies, in their various environments, have roughly analogous and comparable patterns of economics, social organisation, art, and so forth.

Marxist writers in the Soviet Union and the Chinese People's Republic make a combination of approaches and draw China into world history both by stressing concrete interaction and through a world periodisation by 'socio-economic formations'— feudalism, capitalism, etc. Complete isolation, they point out, never existed, although in general interdependence has grown through the ages and only with capitalism do production and exchange become international and is there international division of labour and an international market.[65] Periodisation by socio-economic stage is a manoeuvre formally resembling Hegel's world-spirit periodisation in that it does not require one to solve the problem of independent invention versus diffusion in the successive stages. The orthodox picture between 1930 and 1960, which largely still prevails (but see chapter 4 on the Asiatic Mode), was as follows. World history subdivides into five periods, each corresponding to one of the five prime socio-economic formations. Each period begins when a new formation appears in one or another part of the world. Thus primitive communism, reaching back to the beginning of human society, characterises world history until the time when the first breakthrough to the more advanced slave-holding order occurs in Egypt, Mesopotamia, and elsewhere. The next stage, feudalism, arises first in China (or in Europe and Asia both, but at a later time). Capitalism begins with the bourgeois revolutions of Western Europe and North America from the sixteenth or seventeenth century on (the Dutch or the English revolution). Finally a new period begins with the October Revolution in the Soviet Union, the period of socialism.[66] These stages derive primarily from the study of Europe and are based especially on the transition from feudalism to capitalism; their application elsewhere is somewhat Procrustean though without the Europocentrism of most 'world' history.[67] If the schema is taken too literally, despite all disclaimers, it suggests a thorough weeding out of history so that significant creativity at any period is seen as restricted to a single region, and it conveys a sense that history *must* run in a certain narrow track and no other—a sense false to much ordinary experience and crippling to the imagination with respect to future possibilities. But the arrangement has room for the insights of all three approaches to world history discussed above, and it is based on a consistent cultural materialism that puts the primary focus of cultural evolution in the complex of tools, energy sources, and material environment involved in a society's basic productive activities.[68] In intention it is thoroughly ecumenical and adopts the standpoint of the working majority of mankind, though it raises problems when applied to China, as will be seen.

Notes to this chapter are on pages 129–32

4

WATERWORKS AND
ASSOCIATED IDEAS

The themes of this chapter rest on the older ideas of excess, uniformity, stagnation, isolation and environmental determinism or possibilism, and hold in essence that Chinese society from early times accommodated itself too well to a defective natural setting and thereby lost the possibility of safely generating industrial capitalism (and/or socialism) without outside intervention.

One of the most influential recent summaries is undoubtedly John K. Fairbank's, which he gives under the heading 'China as an "Oriental" Society'. Such societies 'were organized under centralized monolithic governments in which the bureaucracy was dominant in almost all aspects of large-scale activity—administrative, military, religious, and economic—so that no sanction for private enterprise ever became established . . . these governments' customary monopoly of large-scale economic activity will seem to modern Americans most different and "unnatural" '. Geography and waterworks underlay this un-American situation:

Control of the water supply for agricultural production was a strategic factor in the growth of the government's economic function. It was typical of these societies that, unlike Western Europe, they were in regions of semi-aridity where the water supply of great rivers, if properly used, would supplement the insufficiency of rainfall. But irrigation to be effective must be centrally controlled. In like manner its cognate principle, flood prevention, requires centralized direction.[1]

68

Or, in the context of an older essay by a geographer:

> [The proto-Chinese] had developed a culture pattern too rigid and inelastic to permit of progress beyond a certain point. This pattern had adjusted itself closely to its environment; but in so doing it had hardened into a routine from which escape was possible only through the aid of external stimuli. Such a cultural phenomenon has occurred time after time in the world's history. It is being repeated today in China itself on an unprecedented scale.[2]

Cressey, who cites this passage,[3] writes of China's 'cheerful peasants' living in a 'biophysical unity' with nature in this 'old, old land'.[4] 'Where people live so close to nature as in China, an appreciation of geography is fundamental in understanding human affairs.'[5] It is Cressey's desire 'to stress the importance of the environmental restrictions which envelop Chinese life';[6] and he describes China as an 'old stabilized civilization which utilizes the resources of nature to the limit. Until new external forces stimulate change', he adds, 'there is but little internal readjustment'.[7] To this most famous of American geographers of China, the Chinese environment (in implicit contrast to Europe and America) is clearly not 'possibilist', at least for the native Chinese, who are so enmeshed in it that social change for them is a thing of the distant past except as newly effected by western imperialism.

But the most elaborate and determined articulation of this kind of theory comes not from a geographer but from a political and social theorist, Karl August Wittfogel, and it is to his thought and its background that most of this chapter will be devoted. Congruent with Wittfogel but having different emphases are the ideas of Chi Ch'ao-ting and Owen Lattimore at which we will look more briefly.

In these theories too, just as in the conceptions of world history, the main final point is implicitly an active one. The message is when, at the end of his imaginary dialogue with the author, the reader's mind turns to how participation in the social processes of his own world may be made consistent with his new understanding. Where is the social process in China going, as it impinges on us, and what should we do about it? What is possible, where do our interests lie, and who is 'we'? The message is obvious in a far-away example as when a missionary of twenty-two years in China concluded that 'China can never be reformed from within' and that China's pre-eminent need, for Character, Conscience and righteousness, 'will be met . . . only by Christian civilization';[8] it is less obvious but no less present in the geographic theories treated here.

The waterworks theory cannot well be understood apart from the context of Marxian thought on geographic environment and on the Asiatic mode of production.

Marx himself put little stress on the role of geographic environment in social development, although he refers to it in a few passages, as when he writes:

> Not the tropical climate with its over-luxuriant vegetation, but the temperate zone is the mother land of capital. It is not the absolute fertility of the soil, but its differentiation, the multiplicity of its natural products, that forms the natural basis of the social division of labor and by the changes of the natural conditions within which he dwells, spurs man on to multiply his own needs, capabilities, and means and mode of labor.[9]

Engels's conception, I think, has never really been improved upon:

> In short, the animal merely *uses* its environment, and brings about changes in it simply by his presence; man by his changes makes it serve his ends, *masters* it. This is the final, essential distinction between man and other animals, and once again it is labour that brings about this distinction.
>
> Let us not, however, flatter ourselves overmuch on account of our human victories over nature. For each such victory nature takes its revenge on us. Each victory, it is true, in the first place brings about the results we expected, but in the second and third places it has quite different, unforeseen effects which only too often cancel the first . . . Thus at every step we are reminded that we by no means rule over nature like a conqueror over a foreign people, like someone standing

outside nature—but that we, with flesh, blood and brain, belong to nature, and exist in its midst, and that all our mastery of it consists in the fact that we have the advantage over all other creatures of being able to learn its laws and apply them correctly . . .

But the more this progresses the more will men not only feel but also know their oneness with nature, and the more impossible will become the senseless and unnatural idea of a contrast between mind and matter, man and nature, soul and body, such as arose after the decline of classical antiquity in Europe and obtained its highest elaboration in Christianity.[10]

But for the most part Marxists have kept the traditional western division between man and nature, and their distinctiveness is in the emphasis on the intervening economic processes and social relations. If the analytical separation of man from nature is placed in the foreground it will seem (to Marxists and non-Marxists alike) gratuitous to postulate exact equality in the interaction, and the tendency is to lean either toward environmental voluntarism, which gives man the primacy, or toward environmental determinism, in which social development is predetermined by geography. Maoism includes environmental as well as social voluntarism; Plekhanov and following him Wittfogel (in his early works) favour environmental determinism. Plekhanov's strongest formulation, according to Gustav Wetter, is this:

The peculiarities of the geographical environment determine the evolution of the forces of production, and this, in its turn, determines the development of economic forces and, therefore, the development of all other social relations.[11]

And Wittfogel concludes his article on 'the natural causes of economic history' with the words:

Active social labor must take place at every step in dependence on the laws of motion of the objective world external to man—nature. All attempts of whatever kind to shift the primacy to the movements of the social sphere lead back into the realm of free will, into the realm of a freedom of will detached from the hard determinations of the real world around us. Such a path, which priest and mystic take and eloquently recommend for imitation, cannot be trod by true science.[12]

71

But, as we shall see, Wittfogel later did find a need for free will after all.

For Moscow-oriented Communists this question like so many others was settled for about thirty years by a few words from Stalin. He wrote in 1938:

> Geographical environment is unquestionably one of the constant and indispensable conditions of development of society, and, of course, influences the development of society, accelerates or retards its development. But its influence is not the *determining* influence, inasmuch as the changes and development of society proceed at an incomparably faster rate than the changes and development of geographical environment.[13]

In recent years some Russian geographers have got unstuck from the ultimately insoluble dilemma of specifying the relations between man and nature (to which we shall return in this chapter and the next) and have more realistically argued that the geographical environment is not just external 'nature' but a whole comprising both natural and social (productive) processes, and can be adequately studied only by an integrated geography which includes both the physical and the economic sides of the field.[14]

The big Marxist socio-economic formations were mentioned in chapter 3. Each is characterised by a particular *mode of production*, a complex concept referring on the one hand to technology and productivity, on the other to the accompanying social relations, particularly the ownership of productive property and the exploitation of one class by another.[15] Besides the five formations referred to above—primitive communism, slaveholding, feudalism, capitalism, socialism—which in the Stalinist orthodoxy were taken as exhausting all possibilities, there is another, and one which has particularly lent itself to a geographic interpretation: Asiatic society, with its own special Asiatic mode of production.

In primitive communism, economically significant property is owned in common, and there are no classes and no exploitation; in the slave-holding society, the exploiting class owns both the means of production and the producers; in feudalism the lord owns the means of production (land) wholly and the producers (serfs) in part; in capitalism, the capitalist owns the means of production but not the workers, whose labour he exploits by hiring them to work for him; in socialism, there is social ownership of the means of production, and no more exploitation.[16] How such relations are to be defined in Asiatic society has been a matter of controversy. There are two essential groups. On the one hand is the village commune, in which co-operation plays an important role, there is no significant class division into exploiters and exploited, and because of the harmonious presence of both agriculture and handicrafts there is relative self-sufficiency and self-containment. On the other hand is some sort of ruling class, more or less equivalent to the government, an aristocracy or bureaucracy or gentry monopolising literacy and subordinate to a despot. This class owns or controls the land, exploits the communes through tax (equivalent to rent) and corvée, and exercises some economic function such as organising constructions on a scale too large to be undertaken by the individual communes (eg, dikes, canals, roads) or carrying on mining or long-distance trade. Despite long controversy the concept of the Asiatic mode remains excessively vague, and it is not able to supply a functional and progressive class analysis for these societies.

The true issues I think are basically two—they have begun to take shape in earlier chapters. One has to do with the 'Third World' of old and new western colonialism, especially Asia but also Africa and Latin America, to parts of which the idea of the Asiatic mode has also been extended. Are these societies of such different order from European ones that they have no inherent impulse for change in the direction of an ideal 'modern' society —peaceful, wealthy, rational, egalitarian, just? Should they elaborate their own distinctiveness and independence in a context of nationalism and socialism, or must they submit to the

73

guidance of the advanced nations at the price of economic, political, military, and cultural subordination?

The second issue is whether the developed nations (the first two worlds) are reaching a condition like oriental despotism, which can be illuminated and perhaps avoided by the study of Asiatic society—a condition of autocratic rule by a state bureaucracy that directly controls the essential economic functions and through them all other aspects of life: a 1984 or *ta t'ung* world run by a New Class more terrible, by virtue of modern technology, than the old agrarian bureaucracies. These are the capital issues today, and with slightly different reference were already capital in the nineteenth century and in the debates of the late 1920s and early 1930s concerning the Asiatic mode of production.

In Marx and Engels, and indeed all through the history of the Asiatic mode, the progress of Europe has been the real centre of attention. Interest extending beyond Europe—or beyond the contemporary era—has been to fill out a theoretical picture of all types of human society, to see what other times and places can contribute to the most speedy and painless advance for the West, or to seek ways in which the backward lands may transform themselves to resemble western countries, if possible without appropriating all their defects too. Marx and Engels were concerned concretely with the concept in respect of India, China, and Russia. European capitalism, ugly as it was, was unequivocally the instrument of progress in at least the first two of these: thus of India Marx writes that 'its social condition has remained unaltered since its remotest antiquity' until factory-made British textiles came and 'produced the greatest and, to speak the truth, the only *social* revolution ever heard of in Asia'. 'Can mankind fulfill its destiny without a fundamental revolution in the social state of Asia? If not, whatever may have been the crimes of England, she was the unconscious tool of history in bringing about that revolution.'[17] India and China had no progressive social elements—nothing to save; but for a time Marx and Engels thought that Russia might be able to avoid the worst of capitalism by retaining something from the old village commune, the *mir*, to be built into a future socialist society.[18]

Controversy about the Asiatic mode grew in the 1920s when prospects for speedy socialist revolutions in the developed countries of Europe after World War I faded and the interest of the Comintern and the Soviet state in the colonial world, particularly Asia, was quickened. Especially after Lenin's death, the broader (ie, long-term) issues became embroiled in day-to-day decisions and manoeuvring as Stalin struggled with his opposition in Russia and Chiang cut down Communists in China. Behind all the confusion and bad faith, however, the essential question was, what was the most realistic analysis of Chinese society, and who were the allies and the foes of the Communists in their efforts to move the society as rapidly as possible toward socialism, with all due consideration to the international situation—which often was as much as to say, the state interests of the Soviet Union.

The principal objection to the theory was in what it stated or implied about the class situation in contemporary China, where, everyone agreed, there were strong remnants of some pre-capitalist, pre-imperialist order. According to the developing Stalinist orthodoxy, this old order was simply feudal, that is, not wholly independent peasants were exploited by patriarchal landowners who controlled the government and operated it for their own benefit. But in the Asiatic mode, these landowners are more or less equivalent to the 'gentry', whose social position comes not from landownership but from their overlapping and merging with the government bureaucracy; their power comes not from property but from function, from their organising role in respect of production and especially their management of flood control and irrigation. Thus, in contrast to the Leninist conception of government as the *instrument* of the ruling class, the ruling class itself has no definition apart from its virtual identity with the government. Who are the gentry in twentieth-century China, and what does it mean to see in them the remnants of such a system? Marxism-Leninism, focused on Europe, provided no unequivocal answer. If China's first task, in its progress toward modernisation (socialism), is to free itself from leftovers of the Asiatic mode, then Marxism-Leninism is essentially irrelevant; the case is not clearly covered by its theory except in so far as it would suggest that, as in Marx's India, capitalist-imperialists play the leading role, since China was still too backward for the Communists. Questions about the existence of the Asiatic mode in pre-modern times were important primarily as they bore upon these contemporary matters. The 1928 'Resolution on the Agrarian Question' of the 6th National Congress of the Chinese Communist Party did not commit itself on the earlier condition of Chinese society, saying only that the socio-economic system and the village economy *now*, do not *wholly* belong to the Asiatic mode and pointing to several discrepancies, especially the existence of private property in land.[19] In February 1931 a conference on the Asiatic mode was held in Leningrad, and it was made evident that the leadership of the Soviet Union and the International were opposed to the concept: the orthodox line was that there never was a distinct Asiatic mode of production in China or anywhere else.[20] In 1938 Asiatic society was omitted from the socio-economic formations discussed by Stalin in his 'Dialectical and Historical Materialism'.

In the 1950s and especially from 1963 on there has been a revival and a broadening of discussion of the Asiatic mode. Whereas before the principal interest was China, now researchers are thinking about most of the third or underdeveloped world—Asia, the Middle East, Africa, and Latin America. Before, the focus was on class analysis and especially on how to conceive the gentry/bureaucracy; now, the concern is with the self-sufficient, enduring village commune. As before, though, the fundamental question remains, can the concept of the Asiatic mode provide a fruitful means of thinking and deciding about the most desirable steps for these societies toward modernisation (socialism)? And especially—as Marx wondered about the Russian *mir*—can traditional co-operative village practices and de-emphasis of private property in land be carried in some form through to socialism without the divisive and destructive interlude of competitive private property characteristic of feudalism and capitalism? And if so, does this mean that the class of factory workers in these countries is not the vanguard of advance, that the Marxist proletariat is finally the key to progress only in European societies, and traditional Marxism-Leninism is ultimately inapplicable to the rest of the world? And does such an idea imply a denigrating or antagonistic attitude toward the Third World? China, despite the fact that its own revolution was peasant, not proletarian, still rejects the Asiatic mode, and is content with the orthodox definition of its historical stages as slaveholding followed by feudalism, during which there were some signs that capitalism was on the way, then in the last hundred-odd years with the coming of the western imperialists, a semi-feudal, semi-colonial order.[21]

Another difference between the debates of the twenties and thirties and those of the sixties is the de-emphasis, for the time being at least, of what had been a strong minor theme: the geographic conditions of the Asiatic mode, our principal interest in this chapter. In his 1853 article on India, Marx underlined the role of geography:

Climate and territorial conditions, especially the vast tracts of desert, extending from the Sahara, through Arabia, Persia, India, and Tartary, to the most elevated Asiatic highlands, constituted artificial irrigation by canals and waterworks the basis of Oriental agriculture . . . This prime necessity of an economical and common use of water, which, in the Occident, drove private enterprise to voluntary association, as in Flanders and Italy, necessitated in the Orient, where civilization was too low and the territorial extent too vast to call into life voluntary association, the interference of the centralizing power of government, Hence an economic function devolved upon all Asiatic government:

the function of providing public works. This artificial fertilization of the soil, dependent on a central government and immediately decaying with the neglect of irrigation and drainage, explains the otherwise strange fact we now find: whole territories barren and desert that were once brilliantly cultivated, as Palmyra, Petra, the ruins of Yemen, and large provinces of Egypt, Persia, and Hindustan; it also explains how a single war of devastation has been able to depopulate a country for centuries, and to strip it of all its civilization.[22]

As a rule I think those who, like Wittfogel, favoured the concept of the Asiatic mode in the earlier debates also used geography in their social explanations. Environmental conditions seem to go better with the question of bureaucracy and despotism, stressed in the earlier discussions, than with the notion of village commune which attracts more attention now. And, as has frequently been pointed out,[23] the Chinese environment is equivocal in terms of Marx's thought—unlike, for example, Egypt, China has the rainfall to support agriculture without waterworks, but irrigation and flood control are valuable too.

WITTFOGEL

The idea that what westerners saw as the peculiar characteristics of Chinese society were connected with water management, especially the control of the Yellow River, was fairly widespread by the beginning of the twentieth century, but not worked out in full detail.[24] For example, in 1911 the American geographer Ellen Churchill Semple speculated briefly on the social effects of water control in Egypt, Mesopotamia, and ancient North and South America, and of China she wrote:

> It is highly probable that the communal work involved in the construction of dikes and canals for the control of the Hoangho floods cemented the Chinese nationality of that vast lowland plain, and supplied the cohesive force that developed here at a very remote period a regularly organized state and an advancing civilization.[25]

Wittfogel's strategy has been to examine China in the light of Marx's statement (quoted above) about the public works function of Asiatic governments in the dry lands of the Old World. He sees in the statement two 'preconditions for the formation of specifically "Asiatic" forms of society':

1. Agriculture on the basis of *artificial irrigation*.
2. Such a large *areal extent* of the irrigation that is to be carried out in unitary manner, that it cannot be mastered by voluntary effort but requires the intervention of the government in the form of 'public works'.[26]

His lifework has been to develop and extend the consequences of his own conclusion that China does indeed fit this model. He wrote in 1926 that according to Marx those who were in charge of water control in Egypt and India were by that fact the ruling class; and, 'The development of China took the same path . . . Defense against floods and distribution of water [from the Yellow and Yangtze Rivers] . . . over the fields created from early times in China, exactly as in Egypt and in India, a powerful waterworks bureaucracy.'[27] China is essentially a kind of Egypt—an excessively big river in an excessively dry climate, where one lives only at the price of bowing to despotic government. Central planning and centrally administered corvée labour are required to cope with large-area irrigation. Planning, record-keeping, census-taking, and so forth, meant organised control over most of the population as well as over water, the prime production variable; the bureaucracy holding these powers could turn them also to other purposes such as military conquest or the building of roads, great walls, and palaces. By its domination of the economy, the government could also monopolise political power, and China was therefore a single-centred society, a despotism. Power corrupts, and the bureaucrats (the ruling class) manipulated the rest of the population to satisfy their own thirst for power and wealth; the society was permeated with suspicion and terror. The techniques and organisation of this oriental despotism are so potent that even though there is only a relatively small hydraulic core they can be decisive over huge marginal areas—the most notable example being Russia, to which oriental despotism was transplanted from China by the Mongols.

The theory has many dimensions, only a few of which will concern us here. At a minimum we must require of it a faithful rendering of the concrete facts of historical geography; to study this, we need to know when and where the crucial transactions between Chinese society and Chinese landscape took place— when and where did China turn into an oriental despotism? In Wittfogel's work of 1931, *Wirtschaft und Gesellschaft Chinas*, which remains his basic and most detailed statement on China, it is an expansion of Chinese civilisation out on to the Great Plain of north China that he represents as decisive. It will be necessary to quote at some length from this work, which may not be easily accessible to the reader.

> The centres of Chinese agriculture up to the present time have been located, with a 'partiality' wholly incomprehensible to the outsider, quite preponderantly in the area of China's great watercourses. In these places, from the time of 'Yü' on, a large-scale system of protective dikes as well as a similarly extensive system of installations for drainage and irrigation were required. Tasks of this kind and dimension by their nature removed themselves from the control of individual persons, families or villages. Only planned, co-ordinated *mass* labour, *co-opera-tion*, could lay the groundwork for the agricultural work process in the Chinese river regions.[28]

> During the rule of the three dynasties Hsia, Shang and Chou, that is, from about 2200 to 221 B.C., the agriculturally superior and militarily concentrated energy of the river valley dwellers . . . flows forth from the regions of their origin and down the Yellow River over the northern part of the Great Plain, slowly subjecting this extremely attractive but at the same time extremely resistant region to an agricultural tam-ing . . .[29]

> [The hydraulic complex of production] began at that time, after 'Yü', i.e. after the end of the 3rd millennium B.C., to grow out on to the Great Plain. Now waterworks became of decisive importance for the material basis of the expanding empire. But this basis at first developed forms of waterworks of a thoroughly one-sided kind. What was first 'opened up' was the northeastern regions, and that means regions in which there was an eccentric relation between drainage [which term

includes protective works] and irrigation, such, namely, that water *supply* still took place predominantly in small forms (wells), while water *protection* installations expanded to large forms, of hugest dimension. Our thesis, accordingly, is that in the *second* phase of development of Chinese agrarian society, which we put as up to near the end of Chou times, this society entered into a radically new relationship to waterworks, such that in this period *protective* hydraulic installations, at least as far as their areal extent is concerned, came one-sidedly into prominence.

To demonstrate the correctness of this thesis it is not a matter of bringing proof that in the northeast regions great protective waterworks were carried out in order to make the land usable for agriculture— there is general agreement as to this fact . . .[30]

. . . in the epoch from Yü until the end of the Chou period all important hydraulic *protective* tasks, and in part also hydraulic *supply* tasks, by virtue of their great areal extent, caused the intervention of the centralizing power of government.[31]

In another basic statement of 1935, similarly, he writes:

The subjugation of the northern part of this plain, formerly uninhabitable because of the ever-changing loess streams, would have been impossible without the construction of river dams of great magnitude.[32]

. . . the development of public works in the northeast (at first mostly dykes) led to an irrigational agriculture supported by public canal construction, which eventually rose to be the ruling form of agricultural production.[33]

This decisive area, then, is the low and poorly drained parts of the Great Plain lying in today's Hopei, Shantung, and perhaps northern Honan, generally between the higher ground of the T'ai-shan and Lu-shan in the south, the Yen-shan in the north, and the T'ai-hang-shan in the west. And the crucial process of 'growing out' on to the plain took place in the perhaps two millennia between the legendary Yü and the end of the Chou period in 221 BC, especially in the late Chou.

But this picture of the time and place of the formation of Chinese oriental despotism is blurred by other statements. The 'period of China's Oriental Society, i.e., the "empire" ', or 'the

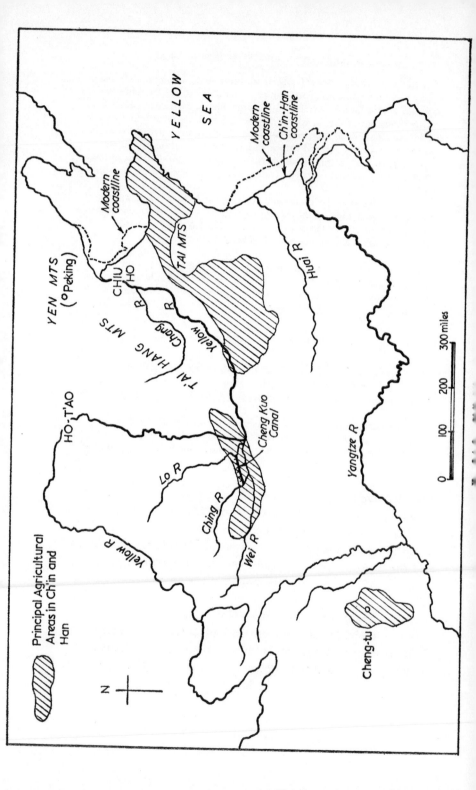

Principal Agricultural Areas in Ch'in and Han

YELLOW SEA

Modern coastline
Ch'in-Han coastline

YEN MTS (°Peking)

CHIU HO

TAI MTS

T'AI HANG MTS

Yellow R

Chang R

HO-T'AO

Lo R

Ching R

Cheng Kuo Canal

Wei R

Yellow R

Huai R

Yangtze R

Cheng-tu

N

0   100   200   300 miles

first epoch of a unified Oriental absolutism' began only in 221 BC, that is, with the Ch'in unification;[34] the 'final establishment of "oriental" society' happened 'during the T'ang period',[35] ie, a millennium later; and there are indications of 'the rise of a hydraulic way of life long before the Shang dynasty'[36] as well as during it: 'Since the climate at that time was semi-arid, as it is today, and since the Shang state was located in the North Chinese lowlands where the erection of large dikes is necessary for permanent living, the Shang rulers must have engaged in extensive hydraulic activities.'[37] A similar ambiguity arises also about the locale, with hints of the importance of humid south China. In *Wirtschaft und Gesellschaft* the spread of Chinese civilisation to the south is not a critical or necessary part of China's configuration as an Asiatic society; yet there are suggestions that the irrigation function of the state is actually better developed in the south than in the north.

> The third epoch of the history of settlement in China is characterized by the inclusion of the rice regions of middle, and later south, China. After the early period, with its local forms of waterworks and the corresponding effects of mere 'colouration' on the function and form of the state, after the second epoch with its huge defensive structures against the flooding tricks of the northern loess streams, we find a double task set to the state in the third phase: in the north the requirements of the second epoch continue since the mechanism of the excessive rising of the Huang Ho and its brother streams remains in force; in the rice regions on the other hand the state must now create and maintain at once drainage [including protective] and *irrigation* works . . .[38]

And again:

> The growth of the Chinese state within the Yangtze Valley compelled it to undertake public works on a new scale. The great rice districts situated around the Yangtze required a very extensive state activity in the construction of waterworks.[39]

The Grand Canal, linking the lower Yangtze region with the older northern centres, is an 'artificial Nile'. Later Wittfogel relaxed his fidelity to Marx and expanded the environmental situation on which the peculiarities of hydraulic society partly rest to include places like south China: the prime conditions are still 'aridity or semi-aridity and accessible sources of water supply, primarily rivers, which may be utilized to grow rewarding crops, especially cereals, in a water-deficient landscape', but 'a humid area in which edible aquatic plants, especially rice, can be grown is a variant of this environmental pattern'.[40] And in *Oriental Despotism* he even intimates that the old north became *less* dependent on its waterworks with the rise of the south:

> The centres of political direction and hydraulic economy coincided more or less until the first millenium A.D., when the growing fertility of the Yangtze area conflicted with the defence needs of the vital northern border zone. From then on, the seat of the central government shifted back and forth; but the northern region *never ceased to be hydraulic to some extent*, and the northern capitals were ingeniously and hydraulically connected with the main rice areas of Central China through the Grand Canal.[41]

But south China, with its relatively broken terrain, abundant rainfall and multiple watercourses, cannot for any purpose be considered a 'variant' of the basic big-river-in-dry-plain type demanded by Wittfogel's theory, which requires not only that water control be necessary (or very desirable) for production, but also that essential water-control tasks surpass the capabilities of all but the highest level in the organisational hierarchy, namely the state. Unless north China is taken as the critical locus for the formation of Chinese society in its Asiatic mould, the theory makes no sense at all.

This ambiguity, arising probably from the difficulty of identifying the necessary critical stages either in the south or (as we shall see) in the north, undoubtedly had something to do with Wittfogel's progressive backing off from the position of strict environmentalism and determinism represented by the statement quoted on p 71 above. Later his stress shifted to technique, then especially to social and political factors and free choice.[42] As his theory has grown increasingly abstract over the years, the concrete details of real natural environments have faded out and his evaluation of the environment's importance has become increasingly equivocal:

> Contrary to the assumptions of the ecological determinists, the natural factor, though essential, plays a formative role only in combination with a number of equally essential cultural factors, some of which are decidedly non-material.[43]

> Ecological determinism oversimplifies the relation between the natural environment and man's technical and economic activities by claiming that this relation is one-sided (with man passively responding to the natural setting) and necessary. In fact, it involves a two-way process; and the ecological setting more often provides the possibility or probability, rather than the necessity for certain types of action.
> But these differentiations do not eliminate the role of the natural factor. They only limit it. And the ecological approach remains central for the understanding of the 'Orient' in which only agro-managerial and state-directed action can solve the problems posed by the natural environment.[44]

Here he seems finally to have come to an elastic kind of possibilism. But most of his reasoning about China is built up on the more deterministic position, as will appear.

Wittfogel has concurrently extended his theory beyond China and even beyond Egypt and India to a vast range of societies in the New World as well as the Old which share features he regards as hydraulic or Asiatic, concluding finally (much like Montesquieu) that hydraulic society 'seems to have shaped the lives of two thirds of mankind'.[45] But China has been and remains at the heart of his theory; he has never published research of comparable bulk or detail on any other society. Within China, 'it was the semi-arid North which, over a long period of time, constituted the dominant centre of power and cultural advance in Eastern Asia'.[46] As we saw, Wittfogel takes the northern part of the Great Plain as the crucial ecological locus of the orientalisation of Chinese society. Thus the historical

85

geography of this region is decisive. It does not support the theory.

As Wittfogel states, it was the margins of the northern Great Plain that were settled first.

> By no means was the whole Great Plain, or even just its northern part, occupied in a steady advance. No, the western agrarian state gained a footing first on the *margins* of the plain, in the northwest, between the T'ai-hang-shan and the Huang Ho, and in the east, on the elevated slopes and in the forelands of the mountain region of Shantung . . .
>
> The *core* of the northern part of the plain on the other hand, in half-amphibious condition with its huge swamps and the multiply branched arms of the Huang Ho and the other streams of the flat land, was according to Legge's opinion a thoroughly inhospitable region up to Yü . . .
>
> [I]t is quite clear that at the earliest at the time of the 'labours' of Yü . . . or not until after Yü . . . the taming by hydraulic technique of the northern part of the Great Plain had its start in the central regions of the plain.[47]

But the spread of settlement came much later than the period Wittfogel is referring to. Even in the Han dynasty, though there was cultivation on the plain east of the T'ai-hang-shan, it was not a specially important agricultural region. Development proceeded from west to east, reaching the middle of the plain during Han and Wei, and the east not until Sui and T'ang. The most important areas lay to the south on the broad belt of higher ground (natural levees and other alluvium mostly at 50–100 metres above sea level) of the watershed between the Huang and the Huai drainage (especially along the upper courses of the Huai affluents) and along the Chi River.[48] In these areas at least, diking was not a precondition for occupance. And in the lower parts of the plain as well, flooding was of a different order from what it became later, so that the sequence of events is not high annual floods in an area otherwise attractive for settlement, followed by diking, then by settlement. Downstream from the Ho-t'ao (the great bend in the vicinity of 40° N, 111° E), ie, in the catchment area for most of its discharge, the Yellow River basin was not denuded of vegetation to the degree that it has been since the time of the Sung dynasty, but was covered with grass, bushes, scattered trees, and forest in various proportions; the same was true of the Huai affluents and the Chi. The northern Great Plain was a zone of grasses, trees, and thickets, broken up by lakes and marshes and by the numerous distributaries of the Huang and other rivers; thus part of the area, probably from around present Te-hsien in the south to Tientsin and Ho-chien in the north, was called *chiu-ho*, 'Nine Rivers'. Vegetation cover, swamps and lakes diminish the quantity of water delivered to stream courses by increasing the proportions of precipitation returned to the air as evapotranspiration and removed from the surface drainage as ground water. The remaining water is spread over greater areas and delivered to the stream courses over a longer period of time, the flow in the system as a whole being at lower velocities. Thus while a greater proportion of the basin will be wet, flood crests are lower and low water is higher. The fact that the coast at the mouth of the Huang, then at about 39° N, was something like thirty miles inland from where it is now would also probably have made speedier the outflow of flood waters from the lower part of the system. Besides, because of the vegetation cover, erosion went much more slowly and stream channels were less apt to be blocked and diverted by silt. This was the pristine landscape; in mythological terms, it was what Yü created, not what he found before him.[49]

The problem of Yellow River floods, and the building of dikes and canals, emerged only gradually in a dialectic relation with settlement: rather than diking being a precondition for settlement, settlement itself largely made the flood problem as vegetation was cleared, swamps drained and streams confined by the growing population.

The histories record that after Yü's regulation of the [Yellow] River there were no disasters from the River for over 1,600 years [ie, until 602 BC] . . . In ancient times there was plenty of land and few people, and where the force of the water reached, people were relocated to avoid it; besides, there were no dikes and hence no floods from bursting and overflowing—following its natural bent, after a flood the water would normally revert to its old channel . . . In later periods, the more regulation was accomplished, the more severe became the damage done by the River . . .[50]

Coming into the Spring and Autumn period [722–481 BC], the royal government declined, the various states struggled for mastery, drawing frontiers and ruling themselves; in the basin of the Yellow River there were dozens of states each planning for its own convenience, some coveting water benefits, some avoiding water damage, and thus there was ponding, breaching, diking, and blocking. And it must have been at this time that work on the Yellow River dikes originated . . .[51]

After centuries of confused disturbance the central plain was a chessboard, with floods on one side and ditches on the other, spread out longwise and crosswise in all directions, suddenly dismantled or suddenly put in operation, completed only to fail, until all was broken and fragmented and there was no ground untouched; in later generations when the [Yellow] River broke out of its course, it poured out on all sides and always had an old track ready-made to take . . . In later times River disasters were more and worse, and the root of the disease is all latent in this [situation] . . .[52]

Definite records of the Yellow River's breaching dikes and flooding date only from the reign of Han Wen-ti, in 168 BC;[53] the kind of large-scale dike construction that was characteristic of later periods began in the time of Han Ming-ti in AD 69.[54] Any earlier dikes, such as the Chin-t'i ('Metal [Gold] Dike') breached by the flood of 168 BC, must have been rather minor constructions on top of the natural levees. Waterworks of various kinds were indeed constructed from probably the seventh century BC on, but they were primarily canals for transport, only incidentally for irrigation. In a few spots, medium-sized irrigation works were carried out too, most notably on the Ch'eng-tu plain in Szechuan, on the T'ai-hang piedmont where the Chang River comes out on to the plain, and north of the Wei in Shensi where the Cheng-kuo canal connected the Ching and the Lo Rivers.[55]

None of these works individually nor all of them together

loom large enough to support the thesis that they gave Chinese society its decisive cast; nor can it plausibly be argued that large-scale protective waterworks assumed a decisive role from Han on, when in any case many of the institutional foundations of traditional China had already been laid.

But in any event it is not dikes but irrigation that is supposed to be central in the theory of hydraulic society, and here again Wittfogel is equivocal. In north-east China (ie, the northern Great Plain, his crucial locale), as he rightly says, wells are much more important for irrigation than canals. 'This would diminish extraordinarily the significance of large-scale water-works for the whole north China region . . . if the inhabitants, so far as they wish to be in safety of their lives . . . [were not forced] to erect protective diking works of the grandest style.'[56] More recently he has said that 'the productive hydraulic acti-vities are usually supplemented by works of flood control that in certain countries, such as China, may be more comprehensive than the irrigation works proper'.[57] Yet in *Oriental Despotism* it is clear that his theory rests on irrigated agriculture in a water-deficient landscape such that the state could maintain a stranglehold on the economy through control of this key sector, in contrast to the individualistic rainfall agriculture of Europe.[58]

And Wittfogel does indeed argue repeatedly that irrigation was necessary, and did in fact exist, in the northern Great Plain from the earliest times.

The arid and semi-arid settings of North China . . . suggest hydraulic agriculture for the ancient core areas.[59]

The divination texts of Shang suggest a climate that, although slightly warmer and wetter, presented a seasonal pattern of semiaridity similar to that existing today. Manifestly whoever wanted to farm effectively in these regions had to irrigate his crops . . .[60]

In fact only a small proportion of the cultivated land in the north China plain has ever been irrigated (the area has been greatly increased in the last two decades), and the main concentrations of irrigated land are, as they evidently have always been, along the T'ai-hang piedmont.[61] As for the earliest period, a recent thorough review concludes that 'there is no evidence, either archaeological or documentary, that the Shang relied upon irrigation agriculture to a significant extent' and that the Western Chou, too, lacks 'any evidence of large-scale irrigation'.[62] The biggest developments in this region came during the Ming dynasty.[63]

The decisive point in respect of irrigation, however, is not whether it was necessary or whether it existed at all, but whether it required organisation at a super-local, ie, state, level, and here the facts are quite clear: such irrigation as there has been on the plain of north China has been almost entirely on a local scale, especially by wells. Thus a survey of the China International Famine Relief Commission reported in 1931 that, although streams and canals (of local scale) were used in some areas, 'on the whole west Hopei [which has the heaviest concentrations of irrigation in the north China plain] lends itself much better to a system of well irrigation than to one of storage reservoirs and canals'.[64] The fact of irrigation *per se*, and such questions as whether intensive agriculture of rice or other crops in China depends on irrigation, and what percentage of the cultivated area is actually irrigated, are utterly irrelevant to Wittfogel's institutional and cultural hypotheses, which depend on the demonstration that requisite water-control tasks so exceeded the capacities of local areas and groups that a central bureaucratic government sprang up to cope with them, and that most other aspects of Chinese culture aligned themselves accordingly. It is astounding that in the teeth of his own data, which show that irrigation in north China (as in the whole country) was overwhelmingly local in character, he still seems to take the irrigation situation as evidence in favour of his thesis about the nature of the society.[65]

This irrelevant emphasis on irrigation points to an important difficulty in Wittfogel's manner of reasoning. He sets out to test

Marx's ideas about Asiatic government, and since Marx wrote that 'artificial irrigation by canals and waterworks' was 'the basis of Oriental agriculture' Wittfogel feels obliged to show that irrigation is the basis of Chinese agriculture as a step toward determining that Chinese society is 'Oriental'.[66] Similarly in respect of works of drainage and flood-protection he wrote:

> According to Marx the intervention of the 'Asiatic' state power extends not only to the regulation of *irrigation* of great areal extent; *drainage* labours belong here too. This obliges us to give a brief but concrete survey of the existence of such provisions for drainage in the most important river regions of China. It is necessary to know of the presence of these tasks and the manner of their solution; otherwise there are elementary preconditions missing for the judgement as to whether or not China belongs to the Asiatic type of society.[67]

Thus in his preoccupation with the letter of Marx's theory he handles his data mechanically. When he writes that even local irrigation can lend a 'specific coloration' to feudalism, or says that Japan had 'an "Asiatic" tinge (irrigational economy on a small scale)', he must simply be suggesting that so long as they concern water, co-operative operations of *any* size—and not just the full-blown state-managed constructions his theory requires —are suspect as being anti-individualistic and tending to give rise to despotic government.[68] There is at work in his thought an overriding logical realism, a belief that reality finally lies in the theoretical model rather than in the concrete facts of social geography. The same bent appears in the development in the comparative direction whereby he seeks primarily to buttress his model in the face of shortcomings in the empirical reality of China; thus he asserts: 'Critically comparative social analysis, and it alone, discovers what are the *decisive* elements [*die tragenden Momente*] of a given social organism and what are the subordinate ones';[69] but it is of no use to the understanding of China to represent it as a defective example of a Platonic type more purely embodied by, say, ancient Egypt. Although comparative study of societies can certainly sensitise one to various possibilities and suggest tentative hypotheses, one cannot well accept a theory unless, even if posited at first from comparative considerations, it might be just as solidly reconstructed inductively from empirical study of the society actually in question. From this standpoint, Wittfogel's comparative approach cannot compensate for the defects in his handling of China itself. China has no concrete landscape and no concrete historical period that can be pointed to as a credible illustration of anything approaching the pure type and mechanism of 'oriental despotism'.

Wittfogel's reasoning is linear and one-way, particularly in the earlier work upon which his later theorising still depends. The world as he presents it is a strict (not probabilistic) causal chain reaching back at one end to general physical processes and the natural environment which is their resultant, and at the other emerging in specific economies, social organisations, governments, arts, sciences. Each condition in this chain is logically and chronologically prior to the succeeding ones, and there is no retroaction in any important way of later on earlier links. For example:

> What determines the state of China's own temperature? Here we touch the question as to the causes of the specific character of the East Asian climate in general. If we succeed in substantially answering it, we will have in hand the key to the understanding of some of the most important preconditions [*Voraussetzungen*] of the economy (and hence of the society, the state order and the world of ideas) of China.[70]

Similarly, the origin of loess explains its physico-chemical character which in turn determines its economic function.[71] Social thinkers, Wittfogel suggests, who neglect this non-social, natural-science side of things are guilty of superficiality or idealism; at best they are foregoing analysis and limiting themselves to simple description or taxonomy: 'Geographical data are essential more to the analysis than to the presentation of institutional history: it is not always necessary to refer to the root in order to identify the flower.'[72] Just as the local combination of determinate physico-chemical processes necessarily results in a given geographical milieu, the milieu in turn acts quite as inflexibly as cause for the specific characteristics of the society inhabiting it.

Actually the introduction of physical, chemical and biological data and terminology does nothing for the analysis other than to imply groundlessly that the social part of it has the same kind of rigour as natural science. There is no demonstrable way in which knowledge of the causes of climatic conditions in East Asia or of the origin of loess, however interesting in other contexts, can add to the understanding of Chinese society—either subjective empathetic understanding or that carried by a predictive model. And the action of loess, climate and the rest of the natural milieu on society is not analogous to the action of mechanical forces on a billiard ball, whose motion becomes predictable as soon as they are fully described. Geographic environment is general in that it embodies necessary causes of social development but not, even in sum, sufficient ones. A study of the physical geography of China gives an idea, as far as inventive imagination reaches, of some of the general courses open to Chinese society now or in the past—there may be more or less fishing, farming, trading, etc, and these natural and economic conditions will in turn supply some necessary causes of social organisation and ideology. But no matter how exhaustive the physical analysis becomes it will never (even through all intermediate links of technology and economic organisation) generate a corresponding increase in specificity of the social analysis or permit detailed conclusions about quality of life or rate and direction of social change, which are the heart of the matter.

The reasoning suffers also from the semantic bind of trying to carry through consistently the analytic distinction between nature and man.[73] It is clear that what 'nature' finally amounts to is definable only as *everything but man*; and Wittfogel, following a hint from Marx, even adds to 'objective' natural conditions, ie, basically the natural environment, 'subjective' ones such as temperament, race, and nation.[74] If the world is divided into two parts, man and not-man, it is absurd and trivial to say that not-man in sum determines man—especially when 'nature' includes culture and heredity. What is left of man to be acted on by nature? Only an ideal individual subject, like the pure abstract free will whose reality Wittfogel denied.[75]

But later Wittfogel thinks of himself as a 'historian [protagonist?] of human freedom',[76] and he adds to his system the element of free choice at one point and only one point: at the beginning of the 'oriental' development. Nature did not compel after all; the hydraulic environment offered the potential, and it was up to the original human group to decide if it was willing to trade freedom and poverty for slavery and wealth. But once the fateful decision is made, free will vanishes again for members of such a society. History offers no more choices to them or their descendants, and the sole hope of rescue from subjugation to a despotic bureaucracy is intervention from outside. This theoretical manoeuvre justifies a low evaluation of China and other 'oriental' societies: given an original choice, they failed, and since then have accordingly been on the rails of an iron determinism such that there is no hope that they can ever reform by themselves.

In sum, Wittfogel's evaluation of the geographic environment is that for China (as for two-thirds of the human race) it was a trap into which the originators of the civilisation fell, betrayed by their insufficient love of freedom and their weakness for plentiful and regular harvests; their descendants are forever crippled by the consequences of this guilty choice unless saved from without. Clearly the action message of this theory is to recommend and justify intervention—presumably economic, military, political, cultural—on the part of the western minority of the world against societies of this sort, whether the old-fashioned genuine ones whose vestiges still dominate the Third World or the new and even more tyrannical and dangerous versions, the 'industrial apparatus' societies of the socialist countries.[77] 'Ultimately,' reads the spooky conclusion to *Oriental Despotism*, 'the readiness to sacrifice and the willingness to take the calculated risk of alliance against the total enemy depend on the proper evaluation of two simple issues: slavery and freedom.'[78]

CHI CH'AO-TING

In his *Key Economic Areas in Chinese History, as Revealed in the Development of Public Works for Water-control* (London 1936), Chi Ch'ao-ting argues a thesis quite like Wittfogel's though more modest and with some difference of emphasis. He sees a traditional China composed of 'tens of thousands of more or less self-sufficient villages' which tended to fall into a number of broad regions the separation of whose economies was not overcome by regional specialisation and inter-regional trade.

> In the circumstances, the unity or centralization of state power in China could only mean the control of an economic area where agricultural productivity and facilities of transport would make possible the supply of a grain tribute so predominantly superior to that of other areas that any group which controlled this area had the key to the conquest and unity of all China. It is areas of this kind which must be designated as the Key Economic Areas.[79]

Grain surpluses generated by means of public waterworks in such an area could be moved and stored so as to maintain the soldiers and officials that unification required. Although Chi writes as if there were several key economic areas (cf also pp 8, 10), in fact, as his frontispiece map shows, he actually comes up with only two, ie, approximately the valleys of the Yellow and Yangtze Rivers east of 108° E and 111° E respectively. Other than these there are only sub-regions and secondary key areas. Thus he is needlessly general and abstract in his theory, when what he is really analysing is the long-term economic and political development of the Yangtze basin relative to the older northern foci. Besides, he cannot prove his hypothesis about the

95

imposition of unity from key economic areas, above all because in the last thousand years unity has been imposed mostly by conquest dynasties from the north, despite the fact that the Yangtze valley has been a key economic area.

Chi differs from Wittfogel in that, although he can use a term like 'iron law' to describe the alternation of unity and division in Chinese history (xiii), there is no hint of environmental determinism (not even of the 'if . . . then' type);[80] the rulers of China do not succumb to the environment, but actively build up and maintain their key economic areas (p 2—again, writing as if there were many of these areas). Another difference from Wittfogel is that Chi says nothing about any organisational implications of waterworks; he does not see the necessity for super-local undertakings as requiring central bureaucratic, and ultimately despotic, control, even deploring in the Yangtze valley the failure of the Chinese state to carry out rational 'centralized collective planning and administration' (p 138). To Chi, the prime importance of waterworks is to assure a grain surplus for the army and the bureaucracy, as well as for revenue to support the extravagant consumption of the ruling class. But in broad terms he, like Wittfogel, believes that China had one or more geographical core regions whose essential characteristic was high grain production through state waterworks, and that the commanding potency of such a core could force the whole Chinese territory into a single uniform political system.

OWEN LATTIMORE

Finally, Owen Lattimore accepts a traditional China like Wittfogel's and Chi's—one based on intensive irrigated agriculture, made up of relatively self-sufficient contiguous local cells —and sets it off against the non-Chinese landscapes of the oasis, the forest, and especially the steppe.[81] He sees the relations between society and environment as subtle and complex, and

avoiding environmental determinism insists, with a concreteness deriving from his own personal familiarity with the peoples
and places concerned, on the necessity for a social interpretation
of environment.

> The beginnings of history depend on the scope that the environment
> allows to a primitive and weak people. Further development works
> itself out as a complex product of the initial momentum, the degree in
> which the environment stimulates or retards the society, and the degree
> in which the society can free itself of control by the environment and
> establish instead control of the environment.[82]

A society uses its environment according to the possibilities
found in it, and thereby changes both itself and the environment; the changed environment then offers different possibilities to the changed society.

> The reciprocal process continues, with infinite possibilities of variation.
> A society may exhaust its habitual use of its accustomed environment.
> Having done so, it may abandon the environment, or it may turn to
> new uses of the same environment . . . [it may] exploit a different
> environment adjoining or accessible to it, which previously it had
> found impenetrable. Or a society may exhaust the resources of an
> environment for one mode of life and abandon it, but leave it all the
> better adapted for the use of some other society with a different mode
> of life.[83]

In practice, though (especially in *Inner Asian Frontiers*), when
Lattimore writes of the two primary social types of China and
the steppe, he represents the range of possibilities as quite
limited. A basically uniform, not very productive, primitive
society which could make use of different kinds of environment
became differentiated by success in the specialised exploitation
of different environments: the loess valleys and plains of north
China, where irrigated agriculture was spectacularly rewarding,
and the grasslands of Inner Asia, where in most places one
could be a rich shepherd but only a poor farmer. Each group
then tended to expand into environments where its own way of
life could be continued; thus the Chinese could expand in-

definitely to the south, but could not move out on the steppe without ceasing to be Chinese—which happened repeatedly to individuals and groups. There were also marginal, ambiguous environments—especially the north China frontier—whose use at a given time depended above all on the political balance between China and the steppe; and the Great Wall was intended to keep the Chinese borderers from casting their lot with the steppe and turning against China, as well as for the obvious purpose of defence against nomad invasions. China had its dynastic cycle, which interacted with changes on the steppe; but to Lattimore, as to Wittfogel and Chi, China was essentially stagnant, so that once one grasps the pattern that events follow, one can talk as well about the Ch'ing as the Han—a view that appears inadequate to the big changes in each period in most areas of Chinese life, from economy and population to governmental practice, speech, ceramics, and art forms. It is this over-generalisation of the patterns of Chinese history that leads to Lattimore's (and the others') preoccupation with origins: the initial specialised adaptation to a type environment (steppe type or irrigated agriculture type) imparts a fateful bias that these societies are stuck with. Thus although his rhetoric is less stringent than Wittfogel's, and he is certainly more concrete and sympathetic and less bemused by the power of government, still his China remains congruent with Wittfogel's.

Let me synthetically recapitulate the gist of the geographic reasoning treated in this chapter. Chinese society took shape in a dry climate with a big river such that irrigation and flood control were needed on a scale beyond the capacities of local groups; while its base remained a honeycomb of relatively self-sufficient villages, a bureaucracy headed by an emperor arose to execute big waterworks. Out of its core hydraulic region the society extended its control to areas where big waterworks were generally less important—especially the broken humid territory of the Yangtze valley and the South—but where population could be dense and administration in the hydraulic style of the North still carried on. In time, the key economic area shifted away from the old dry hydraulic core with its big river, but the society by then had frozen into its pattern of bureaucratic despotism. Over against the Chinese cores of the North and the South is a second kind of primary environment in the steppe, where a herding economy was more profitable; Chinese who leaked into this area were lost to the political economy of farming and bureaucracy and often became its enemies. The repeated interaction between the societies generated in these two types of environment, neither of which could either incorporate or ignore the other, provided a pseudo-dynamism in the ultimately sterile and stagnant history of the Chinese empire. All in all, China's climate is too dry, its plains and rivers too big to be coped with by free, diverse individuals with Judaeo-Christian souls; and the solution to the problem of living there imposed such an extreme and potent social system that from the moment and the point of its origin it reached out to encompass the whole territory of the empire and the whole of Chinese history, until the coming of the West, in a deadening and rigid uniformity.

Unless one has the museum-keeping or curator approach to such ideas about society, or looks at them solely for their beauty or for a crossword-puzzle-like satisfaction, or as time-passers, marks of status or ornaments of conversation, the source of their interest and vitality is in what they mean for social change in the present, and how one should participate in that change—as Wittfogel, for one, fully realises. But even the detached framer or collector of theories about a far-away and long-ago subject like traditional China, whether he is aware of it or not, gets his categories, values, questions, and sympathies from his own society and his own time—so his detachment is self-deception, unless his work is restricted to technical scholarly manipulations of trivia.

Could China undergo fruitful autogenous change in the past?

And by implication (though the connection is rarely made clear), can it and the rest of the 'underdeveloped world' now become modern without following the guidance of Europe and America? The analysis we have been considering, based on historical geography, purports to equip us to answer 'no' to both these questions, but is in fact quite inadequate to them. Whether or not China changed or was going to change in important ways is largely a function of one's view of things in the twentieth century, not of what data one pulls out of the Chinese landscapes and histories.

If it is not left so vague as to be useless, the argument of a special 'Asian' form of society amounts to writing off any possibility of creativity, innovation, or autogenous progress on the part of the great majority of the world's population including the Chinese, and recommending that they await leadership from the developed countries. At the same time the capitalist countries are indirectly warned of the danger of abridging too much the power of business by strengthening government bureaucracy. Meanwhile Chinese civilisation, past and present, is downgraded and any theoretical justification for China's assuming a leading role in the Third World is undermined.[84]

Geographically, these theories say that China (with most of the rest of the world) is less possibilistic and more explainable by environment than western societies, ie, that there is inherently less freedom in China and always has been. Actually, it makes better sense, if one insists on thinking in these terms, to see the Chinese environment as offering enormous potential which was developed very successfully by the Chinese to support nearly a quarter of the human race in a complex stratified society for several millennia. That is the way the old Confucian writers on geography described their country: as uniquely favoured and rich and full of possibilities which their sages were able to actualise. And today too, it is increasingly clear that the geography of China is providing a perfectly adequate basis for extraordinary progress toward a new kind of society—one where everyone's needs are met, and the state no longer functions as the instrument of a minority ruling class in defence of a given system of stratification.

Notes to this chapter are on pages 132–7

5

NATURAL RESOURCES

Modern western writers on human geography have usually seen nature as resources, ie, the kinds, amounts, and spatial distributions of concrete productions of goods and services made possible for a society by its immediate natural environment. Forests, fisheries, scenic landscapes, waterways, or virtually any other part of accessible nature can thus be thought of as some kind of economic resource. In application to China the outstanding themes are industrial minerals and agricultural land, especially the latter, and here again environment is usually represented as deficient or crippling to the development of Chinese society. One may think also of natural resistances,[1] ie, negative resources, and set up a kind of scale at one end of which are the richly and variously endowed environments available to westerners at home and abroad, a world of opportunity, freedom, and progress, and at the other the poorer limited environments of places like China where resistances set limits to progress in the form of ultimate ceiling, slow rate, or restricted range of possible conformation.

China has not always been seen as poor in resources. The image of a China favoured by nature has roots in the old fabulous and excessive East of Croesus, the Magi, and the Pharaohs, and was furthered by Marco Polo and later writers until toward the end of the eighteenth century. In the great collection of writings on China edited by du Halde in 1735, a prime source for later writers, there is much about China's enormous fertility and productivity, its gold, silver, iron, coal, teeming fish, and medicinal herbs.[2] The American geographer Jedidiah Morse expressed the accepted view when he wrote:

> The soil is either by nature or art fruitful of everything that can minister to the necessities, conveniences or luxuries of life . . . China is so happily situated, and produces such a variety of materials for manufactures, that it may be said to be the native land of industry; and it is exercised with vast art and neatness.[3]

And in the latter part of the nineteenth century, owing above all to the researches of the geographer-geologist Richthofen, statements such as the following were commonplace: 'It is estimated that there is just twenty times as much coal in China as in the whole of Europe.'[4] But with the deterioration of conditions in China after the late eighteenth and especially the mid-nineteenth century, the acceleration of industrial development in Europe and America, and the writings of Malthus and other classical economists, a pessimistic estimate gradually came to prevail. Not aggregate resources alone, but the resource/population ratio was emphasised. Already in du Halde there is the statement:

> Despite this abundance, it is nonetheless true to say, which seems a paradox, that the richest and most flourishing Empire in the world, is in one sense rather poor: the land, however broad and fertile it may be, scarcely suffices to feed its inhabitants: it has even been said that twice as much land would be required to support them in comfort.[5]

Likewise Quesnay praised the wealth and abundance of China yet thought it over-populated, with no inch of cultivable land unused. But to him, as to most thinkers before this period in Europe and also in China, a big population was a mark of success, the sign of good government; and anyway China was not unusual in this respect because 'population always exceeds wealth in both good and bad government . . . There are poor people everywhere'.[6] Malthus thought that China had 'probably the greatest production of food that the soil can possibly afford'; 'it is generally allowed that its wealth has been long stationary, and its soil cultivated nearly to the utmost'.[7]

More than any other man it is the geographer George B. Cressey who has set the tone of modern American judgements on the resource/population question in China. 'In Europe,' he wrote, 'the Malthusian doctrine has been temporarily side-stepped by colonization and the development of industry and foreign trade. These possibilities . . . do not appear to be available for China on any significantly adequate scale.'[8] In respect of minerals, he wrote in 1934:

> It is now clearly evident that China is not highly mineralized, and her world rank is that of a minor nation. The available reserves are such that a great development may take place compared with the present, but there seems little possibility that China will ever rival the industrial areas of Eastern North America or Western Europe.[9]

Later he reiterated that because of the inadequacy of mineral resources 'it appears likely that agriculture must remain the basis of national prosperity', and that China cannot 'ever develop a great industrial society'.[10]

Discoveries in the twenty-five years since the founding of the Chinese People's Republic, in a huge country still very thinly prospected, show that there are good reserves of practically all requisite minerals. Probably the only important deficiencies that still exist are in chromite, nickel and cobalt, with petroleum, copper, lead and zinc so far found in only moderate quantities. A Central Intelligence Agency atlas of 1967 considers China's energy resources 'sufficient to support a major industrialization program', and for minerals and metals reports that 'in general the magnitude and quality of Chinese reserves suitable for economic exploitation compare favourably with those of the United States or the Soviet Union'.[11] Although on a per capita basis resources are not particularly impressive, new discoveries generally keep aggregate figures far ahead of use, and there is no reason to assume that in China any more than in the USA or the USSR mineral reserve shortages will be crucial bottlenecks in the progress of industrialisation. In respect of iron ore, for example, one critical mineral in which China has been considered deficient, Cressey accepted about 1,000 million tons of reserves in 1934, and perhaps something over 2,000 million in 1955; but Wang cites Chinese figures of up to 100,000 million tons of potential reserves, and gives his own conservative judgement: 'Discounting exaggerations, the workable reserve appears to be at least 5,000 million tons, more than ample to support a 50-million-ton steel industry.'[12] Such an annual production figure, while it would put China among the leading handful of steel producers in the world, is still low on a per capita basis; to match US per capita production even at the beginning of World War I, for example, China would have to produce about 250 million tons. But the figure is perhaps three times China's steel production of recent years: environmental 'resistance' is not throttling the steel industry now, and it is gratuitous to suppose that at some point in the near future iron ore from at home or abroad will become prohibitively expensive or unavailable, even if one does not allow for technological change permitting use of leaner ores, as has occurred in the United States. Similar arguments apply to the other industrial minerals: new resources are made available by discovery and technological progress as the old ones are used, nor is there any reason to rule out foreign sources where needed.

More prominent in the American public mind is the idea that China's future will be shaped by a shortage of land resources for producing food. Estimates of cultivated land in China have run in the neighbourhood of 267 million acres, ie, 11 per cent of the total area or one-half to one-third of an acre per capita.[13] The easiest farmland is already in use, and with present techniques it would not be worthwhile to cultivate more than a fraction of the remainder; but the population goes on increasing at a rate of $1\frac{1}{2}$–$2\frac{1}{2}$ per cent per annum. From these facts, geographers and others conclude that China cannot feed itself and cannot mobilise the necessary capital to industrialise. Thus though he puts agriculture at the centre of the economy Cressey is not more sanguine about its prospects than about industrialisation:

> Population presses inexorably on the food supply ... From the viewpoint of geography, the basic facts are the limited extent of good land and the restrictions of soil and climate. Any program of improvement must start from the fact that there is only half an acre of good land per capita.[14]

Crowding on the land to the extent that we find in China, writes the Geographer of the US Department of State,

> obviously is far in excess of that which allows optimum development, including the most efficient utilization of the soil and other resources. In addition to the pressure on all available resources, this type of society is incapable of adequately supporting industrial advance. Opportunities for rural inhabitants to save and provide domestic capital are extremely limited. Handicaps also include the inability of the administration to provide such masses with the advantages and even necessities of an adequate social welfare program.[15]

It is particularly population experts who have drawn further social and political consequences from this line of argument. John S. Aird of the Census Bureau shows that the average square mile of cultivated land in China supports far more people than in India or Pakistan because of better farming methods—but then he contrives to interpret this situation to China's discredit:

> ... the inference is that in China the land carries more people in a more delicate balance of population and resources. The system of organization of agricultural production and the distribution of agricultural products may also be more highly developed in China; if this is so, though the immediate pressure of population on food resources may be no greater in China than in India or Pakistan, the Chinese population may be dependent to a far greater degree on the continuity of the system which makes such intensive settlement possible.[16]

In other words, since the Chinese farm economy is more productive and better organised, it is also more vulnerable to disruption—disaster may come at any moment if the 'delicate balance' is upset. This is nonsense: it is in India, not in China, that people starve.[17]

Irene Taeuber and Warren Thompson have made explicit the political and military implications of this set of ideas and suggested their unhappy utility and logic. Taeuber raises

> . . . this very real question: Can China, vast in space and numbers, with agrarian pressures on her resources, develop industrially and maintain her power without exacting the incalculable human costs of selective exterminations?[18]

> Given the absence of international boundaries, there would be major migrations and population redistributions within the Asian region. The potentialities of the slightly used river deltas of southeast Asia and the fabulous resources of the outer islands of Indonesia would not long remain simply potentialities to be developed at leisure by their present holders.

> Expansionism, militarism, and the associated migrations of peoples have characterized nations midway in their modernization. This which has been true of the European and the Japanese pasts may also be true of the Asian futures.[19]

Thompson predicts an increasing *psychological* sense of population pressure among the Chinese, points to the unused but cultivable lands to the south, and continues:

> It is inconceivable that the growing feeling of personal hardship among the Chinese people will not be used by their leaders to create envy and hatred of the peoples who have attained a higher level of living, and thus to build up personal motives for the popular support of conquest. In fact, if the leaders develop a great hunger for power, it would serve their purposes best to keep the level of living low so that they could play on the personal wants of the people.[20]

> . . . the growing feeling of deprivation is a weapon ready-made for the Chinese Communists both in preparing their own people to expand into neighboring noncommunist areas which are rather sparsely settled and also for encouraging many earnest and well-meaning men as well as many frustrated and ambitious men to undertake the subversion of the more moderate governments now in power.[21]

The following passages, one by a geographer and the other by a political scientist, may further suggest how these ideas have come to be taken for granted.

> It may be that China is one of those unhappy countries doomed to choke on its own vast numbers. If China is thrown into convulsions by the ultimate despair of millions of individuals, and if these individuals become aware of the geographic setting of their misery, what is likely to be their avenue of escape? Northward, through Manchuria, into thinly-populated eastern Siberia? Southward, through southeast Asia, perhaps even into thinly-populated Australia? Westward, through Inner Asia, along the ancient routes of invasion, and even into Europe?[22]

> China's major deficiencies lie in resources and in organization. Its movement against the Indochinese peninsula and the Southeast Asian Extension islands is clearly directed toward the acquisition of the abundant minerals and agricultural wealth of these areas, and toward a commanding position in Asia, the Pacific, the Middle East, and Africa.[23]

Such theories, with all their geographic, economic, and political crudity, supply the colour of scholarship and expertise to American prejudices and aspire to provide a rationale for the US presence in East and South Asia. Like oriental despotism, they make it easy to believe that war with China is inevitable and justified. The appearance of necessity depends largely on the physical metaphor of population 'pressure' on resources, especially land. Pressure suggests the weight of a mass of people piled up so that lateral extrusion results, or, still more, a gas increasingly compressed in a closed vessel which can be expected, with all the certainty of a mechanical process, eventually to explode into surrounding spaces of lower pressure. The component particles are practically infinite in number, mindless, and all alike. This image, reminiscent of the old notions of Asia's excess and uniformity, tends to dehumanise the Chinese beyond any sympathetic identification or solidarity with any part of them. They behave without human reason as we know it, out of pure irrational lust for power. Irrationality is necessary to the argument that population pressure and land shortage are driving China toward expansion, because the idea that emigration could provide any sort of rational solution to China's development problems is quite untenable as even Taeuber and Thompson recognise.[24] The annual increase of China's population is the size of the whole population of Taiwan or of Australia, and it is clear that to recruit, support, transport, equip, and house emigrants to even one-tenth of this number, so as just to slow population growth, would be prohibitively expensive and a waste of investment resources that could be much more productively used at home. Thus Thompson is reduced to speaking of an irrational *feeling* of expansionist pressure that China's leaders will manipulate to serve their own obsession with a personal power that is ultimately destructive to themselves and their nation. I suspect that those who have this view were terribly impressed by the destructive irrationality of Hitlerism and make unwarranted extrapolations from experiences of the thirties and forties. Whatever the reason, these theories—population pressure on the one hand, and the irrationality of the leaders on the other—are mainstays of the official American policy assumption that China is aggressive and dangerous, and a principal justification for the high level of US military activity and expenditure in the last twenty years.

Maoist thought handles the resource/population question, much more reasonably and humanely in my opinion, by making an optimistic assessment of resources while at the same time minimising geographic determinism and stressing the human—organisational and psychological—side of development. China has also for the most part followed the policy of slowing population growth through distribution of contraceptives, legalisation of abortion, and promotion of late marriages and small families. In the whole question, Maoism tries to take a 'mass line', making the poor and working people the subjects rather than the objects of policy.

Writers in the Chinese People's Republic describe their

country as having varied climates suitable for all kinds of agricultural production, a cultivated area that could still be more than doubled (although the best land is already in use), good supplies of minerals, and an abundance of waters, fishing grounds, and pastures—a resource situation favourable for peaceful development as it was for the protracted war with Japan.[25]

But Mao, like Stalin, condemns the 'metaphysical or vulgar evolutionist world outlook' that ascribes 'the causes of social development to factors external to society, such as geography and climate'. 'Long dominated by feudalism, China has undergone great changes in the last hundred years and is now changing in the direction of a new China, liberated and free, and yet no change has occurred in her geography and climate.'[26] Why is China's population so strongly concentrated in the southeastern quadrant of the country? Because in the past the social system was unable to cope with any but the easiest environments.

> While natural conditions were partly responsible for the wide disparity in population density, the decisive factor was the social systems which formerly prevailed. While feudal rule squeezed as much as it could out of the impoverished country and imperialism joined in the plunder, there was no way of improving the productive forces; society could not be organized to defeat the aridity of the north-west, to penetrate the forests, dig into the minerals hidden in the earth, open up farmlands, expand cattle raising. Man, therefore, clung to the southeastern parts which enjoy nature's favors.[27]

People are not just passive consumers, they are producers and inventors. Development is frustrated not by lack of resources or techniques, but by poor organisation and ideas.

> The most important thing is that people do not only have a mouth, they also have a pair of hands that can create material wealth . . . People are first of all producers, creators of material wealth. Under present conditions of productive force, human beings can produce more than the consumption materials needed for life, so that human material and cultural life can be progressively raised.[28]

The answer to the population/resources problem lies in revolutionary social change which alone can open up the productive potential of the people and the environment. Mao criticises the idea expressed by Dean Acheson in his 'Letter of Transmittal' to the American White Paper on China[29] that Chinese governments fail and will presumably keep on failing because of inability to solve the food problem created by the 'unbearable pressure' of population on the land. On the contrary, says Mao, 'It is a very good thing that China has a big population . . .'; and 'revolution plus production can solve the problem of feeding the population'.[30]

> Of all things in the world, people are the most precious. Under the leadership of the Communist Party, as long as there are people, every kind of miracle can be performed . . . We believe that revolution can change everything, and that before long there will arise a new China with a big population and a great wealth of products, where life will be abundant and culture will flourish.[31]

Not socialist China, but the United States and the other capitalist countries are relatively overpopulated in that there exist in them starvation and unemployment; because of their social system, these countries cannot effectively mobilise their productive potential for the benefit of their people.[32] 'Overpopulation' if it is to have any economic concreteness means that the poorest people are superfluous, and the 'we', the subject society, would be better off without them. But in Maoism, it is precisely the poor masses first of all that must comprise the 'we' group.

> Some people . . . act as though the fewer the people, the smaller the circle, the better . . . Our large population is a good thing, but of course it also involves certain difficulties . . . Whatever the problem—whether it concerns food, natural calamities, employment, education, the intellectuals, the united front of all patriotic forces, the minority nationalities, or anything else—we must always proceed from the standpoint of overall planning which takes the whole people into consideration . . . On no account should we complain that there are too many people, that they are backward, that things are troublesome and hard to handle, and so shut the problems out.[33]

III

It was in this same context that Mao made probably his strongest statement about the need to slow the growth of population, saying that it should be kept 'for a long time at a stable level, say, of 600,000,000'.[34] A vigorous campaign to limit population growth was in fact undertaken for about six months in 1957, and reduction of births was stressed again from 1962 on. In January 1964 Chou En-lai said: 'Our present target is to reduce the population increase to below 2 per cent; for the future we aim at an even lower rate.'[35] But population is kept in perspective as just one thing among all the many changing elements of Chinese society and economy.

Society is engaged in a war with nature, a struggle for production, and this struggle is like actual war. Within the broad limits of the objective material and social situation, subjective factors—will, initiative, correct decision-making—are what really count; 'it is people, not things, that are decisive'. To revert to the terms I used above, environment can add up only to a necessary cause, and human dynamism must supply the sufficient cause.

> . . . whatever is done has to be done by human beings; protracted war and final victory will not come about without human action. For such action to be effective there must be people who derive ideas, principles or views from the objective facts, and put forward plans, directives, policies, strategies and tactics. Ideas, etc. are subjective, while deeds or actions are the subjective translated into the objective, but both represent the dynamic role peculiar to human beings. We term this kind of dynamic role 'man's conscious dynamic role', and it is a characteristic that distinguishes man from all other beings.[36]

Thus 'there is only unproductive thought, there are no unproductive regions. There are only poor methods for cultivating the land, there is no such thing as poor land. Provided only that people manifest in full measure their subjective capacities for action, it is possible to modify natural conditions'.[37] A poem by Mao contains a pair of frequently quoted lines about the Monkey King Sun Wu-k'ung, hero of the Chinese folk-novel *Pilgrimage to the West*:

The Golden Monkey wrathfully swung his massive cudgel,
And the jade-like firmament was cleared of dust.[38]

Monkey is a symbol of concentrated will, determination and
energy, before which natural and human obstacles fall away.
His magical Taoist mastery of nature and his physical invulner-
ability are equivalent to a complete science and technology;
and in the story itself, he would never have acquired these
powers without the other side of his character, irreverence and
rebelliousness toward established authority whether of heaven,
earth or the underworld. Thus revolution and production go
hand in hand and require the same spirit, which at its most
explicit is a military one.

Westerners too are increasingly disinclined to ascribe any
general determinative role to natural resources, especially after
an economy has successfully begun its modernisation. The
geographer Norton Ginsburg writes that normally 'the role of
the resource endowment is most important in the earlier stages
of economic development, when it acts as a means for capital
accumulation and an accelerator for economic growth if
abundant, and as a depressant upon that growth if niggardly';[39]
Barnett and Morse say that 'the particular resources with
which one starts increasingly become a matter of indifference'.[40]
The economies of any but perhaps the smallest groups are much
too complicated for any simple reasoning from land or minerals
to economic and social change.

In any event, the very concept of natural resources/resistances
is unsatisfactory because it depends on the inelegant metaphysical distinction between man and nature and gives unjusti-
fied emphasis to some parts of an economy while underrating
others. 'Natural' means freely given by nature (though in fact
most resources require work and investment for their exploitation), ie, already there when people come along. But the same
is true of a great deal of 'unnatural' economic material from
year to year and from generation to generation, as far as a
given group's economy, functioning in the present, is con-

cerned. What concrete difference does it make if a thing is 'natural' or the product of past human capital investment? If there is a lake full of fish, the important questions are what kind they are, what they should be fed, how many may be caught for maximum steady production; whether the first fish were naturally there or whether someone once stocked the lake is irrelevant to their economic use. If there is a strip of level land one wishes to farm, it does not matter to the economy whether it was naturally level or made so by one's forefathers; what counts is whether it is practical to water it, whether corn or wheat would do better, and so forth. The emphasis on natural resources, or 'land' in its narrow or broad sense, implies an archetypal drama of a first group of people in a primeval land-scape, looking round to see what is already there that they can use—what the ancestral emperor Yü laid out for them. But in reality, virtually all human groups have accumulated enor-mous changes, both investments and disinvestments, in their landscapes for thousands of years. It is not the *source* of an ele-ment in the man-landscape complex that is important, but that the element presently functions in that complex. I argued this point above in the context of Wittfogel's thought. From the point of view of material things, 'natural' resources and 'arti-ficial' capital stock are the same: material organised so as to facilitate production—sites, fuels, tools, waters, raw or partly finished materials. From the point of view of myth, 'nature' in China today is the whole material world, with which the whole people struggles by means of social labour and science, in the image of the class struggle.

Thus the simple population/resource analysis is subject to the same criticisms as the equally abstract population-growth/capital-investment model. 'Most of the approaches to the problems of population growth,' writes Gunnar Myrdal, '. . . have usually this in common: the economic effects of a population increase are considered in terms of the man/land ratio and the static law of diminishin returns.' He finds such theories too mechanistic and schematic to be of use in analysing real development problems.

> Economic development means, in the first place, that more people will work, that they will work longer and more efficiently, and that they will cooperate in order to create institutions that make this possible and rewarding . . . The availability of more capital [*scilicet* or resources] can, of course . . . increase labor productivity . . . The same effect can also be expected from many specific policies: for instance, policies directed at effective land and tenancy reforms; at raising levels of nutrition, health, and education; providing a larger and better trained cadre of managerial and supervisory personnel at all levels of responsibility; improving efficiency and honesty in public office; and, generally, promoting a consolidated nation ruled by an informed and determined government.[41]

A real functioning economy consists of a very large number of material things plus concrete pieces of behaviour recurring in patterned ways, and which interact along many paths in systems of multiple cause-and-effect relations. Ores or soils are no different from any other element just because they are 'natural'; there is no special geographer's wisdom that allows short-cut predictions through assessment of 'environment'. The two-factor man/resources or man/land analysis, with the attached impossible condition 'other things being equal', cannot attain even a rough approximation to reality.

As far as concerns farmland, these models assume that food production in China has a definite ceiling that is near at hand if not already reached. Malthus and Cressey, a century apart, both describe China's agriculture as at or near this limit, though in fact the population had doubled in the intervening years and production techniques were still basically the same. The idea of such a ceiling, at least for the short run, is built in to the typical model of agriculture and population, exemplified in its application to China by Arthur G. Ashbrook, Jr. The argument is diminishing returns to labour.

> All that is necessary is to accept the notion that once China's land suitable for agriculture is being busily worked, a doubling of the labor

force will not double output from the land, because each worker's labor now is being applied in effect to only half as much land. Suppose we have in 1957 the following illustrative situation:

    (a) The amount of agricultural land, the level of agricultural technology, and the stock of farm machinery and farm structures are fixed in amount in the short run.

    (b) The population is 600 million; the output of grain . . . is 200 million metric tons . . .

    (c) A genie is found in a bottle by a poor but honest economist on the coast of the Yellow Sea. At the bidding of the economist . . . the genie selects at random 200 million of the Chinese people and makes them disappear.

    (d) The output of grain is reduced . . . only 10 percent [though population dropped by a third. Consumption rises above the subsistence level, and there is a surplus to trade abroad] in return for industrial equipment and cannon. So ends the highly simplified illustration, which does no violence to the basic facts in China.[42]

To assume, as he does in (a), that the factors of agricultural production other than labour are fixed is quite incompatible in the real world with the one-third reduction in population, unless there is nuclear-biological-chemical war—hence the genie in the bottle. Population and agriculture in a peasant economy are intimately related and both are changing complexly and interdependently, not as abstract aggregates. That agricultural production must have a ceiling or a maximum rate of increase insufficient for population size or growth is not at all self-evident. Agricultural production in China or anywhere can be enormously increased by improved fertilisation, return of organic wastes to the soil, irrigation, drainage, insect and weed control, new crops and strains of seed, and so forth. The 'natural' limits are biological and ultimately only physical; they depend on the amount of solar energy available at the earth's surface. These limits are nowhere in sight even in countries where farm labour has much higher productivity than in China.[43] To make much of such limits simply diverts attention from social barriers to increased production and more equitable distribution, and the fact that removal of these barriers has resulted in unprecedented improvement in both during the last twenty-five years.

This is not to say that all resources, whether agricultural or mineral, are ultimately of equal value. The point is that within time spans that have reality for us, China is not peculiarly handicapped relative to any other part of the world in respect of the presence of 'resources' or the possibilities of making better use of them.

> There has been a certain tendency to regard technological advance as a chancy phenomenon, a bit of luck that is sure to run out sooner or later (with the ever-present implication that it will be sooner) . . . The view that improvements must show a diminishing return is implicit in the thought of those who regard more optimistic opinion as 'cornucopian'. Yet a strong case can be made for the view that the cumulation of knowledge and technological progress is automatic and self-reproductive in modern economies, and obeys a law of increasing returns.[44]

Even from a narrowly technological viewpoint (and social and political questions are more important), it is logical to assume in accordance with all modern experience that new advances will continue to be made in China too at an increasing rate, in agriculture as in other areas; and Chinese society is now organised to take good advantage of them.

Notes to this chapter are on pages 138–41

6

CONCLUSION

A broad social approach to geographic thought helps clarify the parts of fact, theory and myth that go into ideas about China, or other areas of the world. On the ordering concepts of region and geographic environment, definable in principle by cool blends of fact and reason, arise theories like environmental determinism, possibilism, population pressure, which merge into mythic designs such as East and West, Asia and Europe, diversity and uniformity, the geographic progress of history— grand formalised world maps fraught with meaning for particular groups. Calling them mythic emphasises that they are story-patterns involving the unifying purposes of their cultures and classes. The serious Chinese and western theories of the geography of China are all more or less consonant with fact and logic: China *is* big, populous, at the eastern extreme of Eurasia, subject to flood and drought, mostly surrounded by regions that sustain much thinner populations, and so on. But these theories exclude each other at the point where the interests they are constructed to serve become contradictory.

The solution is not a pose of detached relativism, nor alone a painstaking study of the facts and logic of each case, but making explicit the purpose of a theory and the reference-group of a myth. I hope this book will help give credence to the axiom that historical contexts and social functions are not extraneous to ideas. A social theorist is not persuasive by fact and logic alone; we must in some degree trust his judgement that he has got his masses of material into proper perspective. But the warmth with which he convinces us to accept his perspective may arise from loyalties he does not state, perhaps not even to himself. It is much preferable to have these out in the open so that we are in a position to evaluate the argument as a whole.

Obviously, not all myths or theories are going to make equal sense. In particular, the idea that the élite of a stratified society deserves—because of birth, virtue, cultural or economic productiveness—a position of power and wealth in its own country or in the world generally, and that its interests are identical with everybody else's (what's good for General Motors is good for the USA), requires some sort of hocus-pocus with facts and logic, whether this is done through geography or some other social science. Hence such obliquities as Bowman's possibilism, Thompson's populationism, Wittfogel's oriental despotism, and so forth, which at the crudest level are at the same time examining such matters as the prospects for using China as a colony, a military scare, or an ally, or the means of showing Chinese socialism to be peculiar, unattractive and irrelevant to the Third World as well as to the USA and the rest of the 'First World'.

If one really tries to take the standpoint of the interests of the bulk of the human race, rather than the interests of the élites, things look rather different. Stratified societies, with their systematic maldistributions of wealth and power, are seen to be unsatisfactory and in need of change. Rather than emphasising static 'nature' and social processes derived from or analogous to it, we will give attention to man's 'unnatural' imaginative dynamism and creativity. China will no longer appear as crippled by a defective natural environment, unable to progress without plenty of First World intervention, but as a most advanced society to be regarded with serious interest and hope. Such new societies had better continue to be possible if the human species is to survive.

Since this manuscript was basically completed there has been some change in US-China relations, the most dramatic moment being President Nixon's visit to China early in 1972. It is no longer fashionable to condemn China with the vehemence of the 1950s and 1960s, and the *New York Times* writes People's Republic of China instead of Communist China. But old attitudes have changed only superficially. The tone is still patronising, the Chinese are still the antipodal people, the soul of Asia is still inscrutable.

> Perhaps no voices are more foreign to an American than those of the Chinese. The two peoples are poles apart—separated by such a vast cultural, historic and psychological chasm that it is hardly surprising that Americans have so much trouble grasping that elusive sense of 'Chineseness'.[1]

We may have a vogue for clever import goods and the subtleties of acupuncture, but the society itself is irrelevant because of this foreignness—and because we are sophisticated and modern, while China represents a primitive stage that we have long left behind. John Fairbank, again, enunciates the new formulas.

> For Americans . . . the Chinese revolution hardly offers a model to follow; its methods don't scratch where we itch. [James] Reston [of the *New York Times*] felt 'constantly reminded here of what American life must have been like on the frontier a century ago . . . This country is engaged in one vast cooperative barn-raising . . . They remind us of our own simpler agrarian past.' This is appealing but carries little message for the American future.[2]

The socialist thought that sums up the purposes and methods of the Chinese revolution is nothing but 'ideology', assumed to be, in China as in the US, a cynical brainwashing trick to uphold the only realistic goals, power and wealth. 'Nixon and Mao were once vociferous ideological opponents,' writes Fairbank, 'so that their meeting gives us a healthy skepticism about ideology in general.'[3]

Clearly, China is very different from the United States in geography, in historical culture, above all in being a peasant society. Fairbank is right that it is not a blueprint for America. But by its organisational forms, its technology, its educational experiments, its public health programme, its social security, its housing projects, its popularisation of culture, its recycling of wastes, and innumerable other novelties, mistakes as well as successes, it is breaking new ground and showing the hopeful possibilities that exist for any society when it once decides that the time has come for its productive resources to be owned by its people and operated in their interest, rather than hogged by a privileged fraction.

Notes to this chapter are on page 141

NOTES

CHAPTER I   INTRODUCTION

1   The term belongs to Étienne Juillard, who uses it pejoratively
    in 'Aux frontières de l'histoire et de la géographie', *Revue
    historique*, 215 (1956), 270, in criticising Fernand Braudel ('La
    géographie face aux sciences humaines', *Annales—Économies,
    Sociétés, Civilisations*, 6 (1951), 485–92).
2   For introduction to the literature, see G. Tatham, 'Environ-
    mentalism and Possibilism', in Griffith Taylor (ed), *Geography
    in the Twentieth Century* (London and New York 1957), pp 128–
    62; Marvin Harris, *The Rise of Anthropological Theory* (New
    York 1968), ch 23; Pitirim Sorokin, *Contemporary Sociological
    Theories* (New York 1964), ch 3; Harold and Margaret Sprout,
    The Ecological Perspective on Human Affairs (Princeton 1965).
3   Harris, op cit, p 662.
4   Man and Nature appear as counterplayers in game theory for
    example in Peter R. Gould's 'Man Against His Environment:
    A Game Theoretic Framework', *Annals of the Association of
    American Geographers*, 53 (1963), 290–7.
5   C. P. FitzGerald, *China: A Short Cultural History* (London 1965),
    p 1.
6   Edmond Demolins, *Comment la route crée le type social* (Paris nd),
    1, ix; quoted in Sprout, op cit, p 51. His discussion of China
    (1, 245–73), which sees steppe types dominating a peasantry
    shaped by the route of Tibet, will not concern us.
7   On imaginary experiment, see Nicholas S. Timasheff, *Socio-
    logical Theory: Its Nature and Growth* (New York 1967), pp 174,
    224; R. M. MacIver, *Social Causation* (New York 1964), pp
    258–9, 264–5.

8 Cf the discussion of Stalin's and Mao's ideas of geographic environment, ch 4 and 5 below.

9 Columbia University, *Spectator*, 29 October 1968.

10 C. E. Black, *The Dynamics of Modernization: A Study in Comparative History* (New York 1967), p 158.

11 Ibid, pp 84, 90–1.

12 Cf Morton H. Fried, *The Evolution of Political Society* (New York 1967), pp 52, 186, 189, 235.

13 Franz Schurmann, 'Chinese Society', in David L. Sills (ed), *International Encyclopedia of the Social Sciences*, 2 (1968), 411–12.

14 G. William Skinner, 'Marketing and Social Structure in Rural China', parts 1, 2, and 3, *Journal of Asian Studies*, 24–5 (1964, 1965).

15 The best presentation of the older Chinese geographic literature in English is Joseph Needham, *Science and Civilisation in China*, 3 (Cambridge 1959), 497–590. See also Liang Ch'i-ch'ao, *Chung kuo chin san pai nien hsüeh shu shih*, part 15c, 7, 8 (Taipei 1958), pp 298–324; Wang Yung, *Chung kuo ti li hsüeh shih* (Changsha 1938); idem, *Chung kuo ti li t'u chi ts'ung k'ao* (Shanghai 1956); Chang Ch'i-yün, *Chung kuo ti li hsüeh yen chiu*, 1 (Taipei 1955), 231–81; Wang I-yai, 'Chung kuo ti li hsüeh shih', in Lin Chih-p'ing et al, *Chung kuo k'o hsüeh shih lun chi*, 1 (Taipei 1968), 67–121. On cartography see also Cheng Ch'iao, *T'ung chih lüeh*, ch 48; Chin Yü-fu, *Chung kuo shih hsüeh shih* (Peking 1962), pp 205–12; Chu Shih-chia, *Chang Hsüeh-ch'êng, His Contributions to Chinese Local Historiography* (unpublished PhD dissertation, Columbia University 1950), pp 89–94.

16 Chin Yü-fu, op cit, pp 120–4; on privately written works see pp 205–12.

17 Li Tsung-t'ung, *Chung kuo shih hsüeh shih* (Taipei 1962), pp 102–3; P. Demiéville, 'Chang Hsüeh-ch'eng and his Historiography', in W. G. Beasley and E. G. Pulleyblank (eds), *Historians of China and Japan* (London 1961), pp 167–85, esp p 174.

18 Ch'iu Chün, *Ta hsüeh yen i pu*, 10a.

19 *Chou i*, 7.1a. *Ch'ien* and *k'un* are hexagrams standing for heaven, male, etc and earth, female, etc, respectively.

20 Liu Chih-chi, *Shih t'ung*, 14b; E. G. Pulleyblank, 'Chinese Historical Criticism: Liu Chih-chi and Ssu-ma Kuang', in Beasley and Pulleyblank, op cit, p 147.

21 *Han shu*, 28B.19b. The dictionary *Tz'u hai*, 2, 364, sv *feng su*,

quotes this passage and comments that in paired contrast there is a slight difference between *feng* and *su* but separately they may mean the same.

22 *Ch'i*, 'to present food', but much more common in what were originally loan meanings for the *ch'i* of the *Shuo Wen*'s *yün ch'i*, 'cloudy vapours'; thus air, breath, vapour, temperament, disposition, vital principle—Karlgren, *Grammata Serica Recensa*, 517a, c.

23 *Li chi*, 12.15b. Notes by Cheng Hsüan, Han dynasty.

24 *Hou Han shu*, 54.20b.

25 *Meng tzu*, 13.5b, trans Legge, 2, 331.

26 *Chou i*, 2.11a, trans R. Wilhelm, p 83.

27 *Shang shu*, 18.6b.

28 Tu Yu, *T'ung tien*, 185.985. Cf the similar but less developed statement in the sixth-century work *Liu tzu* by Liu Chou, 46 (*feng su*), pp 58–9.

29 See especially Derk Bodde, *China's First Unifier* (Leiden 1938).

30 Étienne Balazs, *Chinese Civilization and Bureaucracy* (New Haven 1964), pp 10–11.

31 Ibid, p 26.

32 Ibid, pp 24–5.

33 C. Northcote Parkinson, *East and West* (New York 1965), p 248.

34 Bernhard Karlgren, 'The Book of Documents', Ostasiatiska Samlingarna, *Bulletin*, 22 (1950), 33; *Li chi*, 21.2a. Cf Fung and Bodde, 1, 377–8.

35 Laurence G. Thompson, *Ta T'ung Shu: The One-World Philosophy of K'ang Yu-wei* (London 1958); Fung and Bodde, 2, 675–91.

36 Mao Tse-tung, *Hsüan chi*, 4 (Peking 1960), 1474, 1476, 1481; *Selected Works*, 4 (Peking 1967), 412, 414, 418, 423.

37 Erik H. Erikson, *Childhood and Society* (New York 1963), p 345. Granet argues for the social origin of Chinese conception of space and time and the physical world (*Pensée Chinoise*, pp 90, 101, 112, 343 et passim).

CHAPTER 2   THE MYTH OF ASIA

1 Cf William Bunge, *Theoretical Geography* (Lund 1966), pp 14–23. See also Derwent Whittlesey, 'The Regional Concept and the Regional Method', in Preston E. James and Clarence F.

Jones (eds), *American Geography: Inventory and Prospect* (1954), pp 19–68.

2 Roger Minshull, *Regional Geography: Theory and Practice* (London 1967), p 38.

3 For continental division and associations attaching to Asia I have relied especially on the following works, and on those cited hereafter concerning the idea of Europe: C. Raymond Beazley, *The Dawn of Modern Geography* (New York 1949); E. H. Bunbury, *A History of Ancient Geography* (London 1883); *Paulys Realencyclopädie der classischen Altertumswissenschaft*, 2 (Stuttgart 1896), sv 'Asia'; J. Oliver Thomson, *History of Ancient Geography* (Cambridge 1948); H. F. Tozer, *A History of Ancient Geography* (New York 1964); C. van Paassen, *The Classical Tradition of Geography* (Groningen 1957); John Kirtland Wright, *The Geographical Lore of the Time of the Crusades. A Study in the History of Medieval Science and Tradition in Western Europe* (New York 1925).

4 Tozer, op cit, p 67.

5 Orosius, *Historiae adversus paganos*, 1.2 (p 9); Augustine, *De civitate Dei*, 16.17 (col 497), cited below.

6 Augustine, *De civitate Dei*, 16.17 (col 497); similarly Isidore, *Etymologiae*, 14.2.3 (col 496).

7 Denys Hay, *Europe: The Emergence of an Idea* (Edinburgh 1957), p 125.

8 Eg, Augustine, loc cit (after the passage quoted above): why did Egypt not fall to the Assyrians, 'who are said to have held all of Asia except India alone?'; Isidore includes Egypt in his section on Asia, 14.3.27; also Orosius 1.2, p 11.

9 Carlo Curcio, *Europa: storia di un'idea* (Firenze 1958), pp 80–1, 353; Jürgen Fischer, *Oriens—Occidens—Europa* (Wiesbaden 1957), pp 10–19; Hay, op cit, pp 8–15; Isidore 9.2.37 (col 331).

10 Wright, op cit, pp 157–63; Beazley, 1, op cit, 377.

11 Parallels do exist in China in the cosmology of Tsou Yen of the third or fourth century BC (Fung Yu-lan, *A History of Chinese Philosophy*, 1 (Princeton 1952), 160–1), and in the awareness from at least the early centuries AD of Indian civilisation as the source of Buddhism. But none of this ever loomed large enough to upset the identity of the true ecumene with the Chinese civilisation, in the sense described in the last chapter.

12 Herodotus, *The Persian Wars*, 1.4, p 5.

13 Ibid, 7.101–3, pp 534–5. Cf also 7.135, pp 546–7.

14 Ibid, 2.17–19, pp 124–5; 2.35, p 133.
15 Ibid, 3.94, p 259; 3.98–106, pp 260–4.
16 Ibid, 3.107–13, pp 264–6.
17 Ibid, 4.1–117, pp 290–336.
18 *Airs Waters Places*, 12, pp 105–9; cf Herodotus, op cit, 1.142, p 78.
19 *Airs Waters Places*, 16, p 115.
20 Ibid, 23, p 133.
21 *Politics*, 7.7, p 291; cf 3.14, p 157.
22 Curcio, op cit, pp 70ff.
23 Curcio traces them in Marsilio of Padua, Machiavelli, Jean Bodin and others (pp 206–25).
24 Cf Fischer, *Oriens—Occidens—Europa*, pp 60–3, and this whole chapter, 'Die religiöse Bedeutung des Gegensatzes "Oriens-Occidens" ', pp 59–74.
25 Ibid, p 72.
26 Orosius, 2.2.9–10, p 85; Fischer, op cit, pp 26ff.
27 Fischer, p 54.
28 Otto Bischof von Freising, *Chronik oder die Geschichte der zwei Staaten* (Berlin 1960), Prologue to Book 1, p 14; cf Prologue to Book 5, pp 372–5; 5.36, pp 426–9. For other formulations see Curcio, op cit, pp 82 (Augustine), 128–9; Wright, pp 233–5.
29 Bonifatius Fischer, *Vetus Latina: die Reste der Altlateinischen Bibel*, 2: *Genesis* (Freiburg 1954), Genesis 2.8; cf Wright, op cit, pp 261ff; Beazley, op cit, 1, 332–4. Isidore put paradise 'in Orientis partibus' and makes its climate sound like Hippocrates' Asia: 'There is no cold or heat there but ever the mildness of spring' (14.3.2, col 496).
30 Curcio, pp 281–2.
31 Roger Bacon, *Opus Majus*, ed John Henry Bridges, 1 (Oxford 1897), 368; Isidore, 14.3.12 (col 498); Richard Wallach, *Das abendländische Gemeinschaftsbewusstsein im Mittelalter* (Leipzig and Berlin 1928), p 8.
32 Rudolf Wittkower, 'Marvels of the East: A Study in the History of Monsters', *Journal of the Warburg and Courtauld Institutes*, 5 (1942), 159–97; Wright, pp 274–5, 359.
33 For details see eg Isidore, 14.3, 14.6 (col 496ff).
34 Beazley, 1, 335–8; Wright, pp 267–8, 287–8.
35 Wright, pp 283–6.
36 Giraldus Cambrensis, *Topographia hibernica*, 1.34–40, pp 68–73; in the translation by Thomas Forester, 1.26–28, pp 52–6.

37 Henry Yule, *Cathay and the Way Thither* (Taipei 1966), pp 16–17. Originals and translations of texts concerning the Far East are conveniently assembled in Georges Coedès, *Textes d'auteurs grecs et latins relatifs à l'Extrême-Orient* (Paris 1910). See also Wright, pp 271–2.

38 Herodotus, 3.116, p 267; see also 3.106, p 263; cf van Paassen, p 171.

39 2, Prologue, p 75 (trans Forester, p 58).

40 Eg Bacon (op cit, 1, 372) puts it 'in extremitate orientis'. Bacon, incidentally, correctly identified 'Cathay' with the 'Seres', whatever others may have thought: 'magna Cathaia, quae Seres dicitur apud philosophos' (loc cit). Wright (p 271) following Yule (op cit, pp 180–2) says the identity was not recognised until the sixteenth century, which on Yule's evidence seems generally true.

41 Wright, p 357.

42 A great deal has been written about the idea of Europe in recent decades. See especially: Carlo Curcio, *Europa: storia di un'idea* (Firenze 1958); Christopher Dawson, *The Making of Europe: An Introduction to the History of European Unity* (London 1939); Gonzague de Reynold, *Le monde grec et sa pensée* (Fribourg en Suisse 1944); Denis de Rougemont, *The Idea of Europe* (London and New York 1966); Luis Diez del Corral, *El rapto de Europa: una interpretación historica de nuestro tiempo* (Madrid 1954); Jürgen Fischer, *Oriens-Occidens—Europa: Begriff und Gedanke 'Europa' in der späten Antike und im frühen Mittelalter* [fifth to eleventh centuries] (Wiesbaden 1957); Werner Fritzemeyer, *Christenheit und Europa: zur Geschichte des Europäischen Gemeinschaftsgefühls von Dante bis Leibniz*, Beiheft 23 der *Historischen Zeitschrift* (1931); Heinz Gollwitzer, *Europabild und Europagedanke: Beiträge zur deutschen Geistesgeschichte des 18. und 19. Jahrhunderts* (Munich 1964); Denys Hay, *Europe: The Emergence of an Idea* (Edinburgh 1957); Richard Wallach, *Das Abendländische Gemeinschaftsbewusstsein im Mittelalter* (Leipzig and Berlin 1928).

43 Dawson, op cit, p 4.

44 Oscar Halecki, *The Limits and Divisions of European History* (New York 1950), p 13.

45 Diez del Corral, op cit, p 104.

46 Reynold, op cit, p 89.

47 Ibid, p 71.

48 Ibid, p 81.
49 Gollwitzer, op cit, p 20.
50 Dawson, p 8.
51 Ibid, p 106.
52 Wallach, op cit, passim.
53 Dawson, p 135.
54 Fischer, pp 50–1.
55 Fischer, p 113; Fritzemeyer, op cit, p 14; Wallach, pp 26–9.
56 Fischer, p 18.
57 Berthe Widmer (ed), *Enea Silvio Piccolomini* (Basel and Stuttgart 1960), pp 448–50; Wallach, pp 51–2; Fritzemeyer, pp 23–9.
58 Cf Curcio, pp 261–2, 351–2, 446; Gollwitzer, p 39.
59 For Dawson, Europe was apparently 'made' much earlier since his *Making of Europe* stops at the eleventh century. Donald F. Lach's *Asia in the Making of Europe* (Chicago and London 1965) goes up to 1800. Cf Denis de Rougemont, pp 380–1: between the world wars there were two schools of culture historians of which 'One . . . continued the tradition of the Enlightenment . . . and regarded Europe as a creation of the Renaissance' while 'The other . . . regarded the great centuries of the Catholic Middle Ages (from the eleventh to the thirteenth) as the only Europe worthy of the name'.
60 Curcio, p 818; see also p 819 and pp 608–39; Hay, pp 123–5. A really scurrilous write-up of this theme is Parkinson, op cit, ch 20, where it is put in strongly racist terms.
61 Reynold, pp 270–1; cf Hay, p 3.
62 Giraldus Cambrensis, 1.37 (pp 70, 71); Forester 1.27 (pp 54, 55).
63 Hippocrates, as Curcio points out (pp 58–9), makes the Europeans diverse among themselves in contrast to the uniformity of Asia. Strabo (*Geography*, 2.5.26) emphasises the advantages of Europe's geographic diversity.
64 Friedrich Schlegel, 'Über die neuere Geschichte', *Kritische Friedrich-Schlegel-Ausgabe*, 7.1 (Munich 1966), 131, 132–3.
65 John Stuart Mill, *On Liberty* (Chicago 1955), pp 103–6. A modern degeneration of the idea is exhibited in Parkinson, p 248.
66 In Ernest Barker et al (eds), *The European Inheritance*, 3 (Oxford 1954), 310, 311ff.
67 Derwent Whittlesey, *Environmental Foundations of European History* (New York 1949), p 132.

68 P 25. Also Reynold, in the passage quoted on p 36 above, says Byzantium is 'composite', therefore inferior to Greek purity; but the heterogeneity of Greece's origins is praised (pp 68–9).

69 Thus Reynold tells of studying the Persian wars in school: 'We received at that time our first lesson of patriotic piety, of civic and military spirit. Since the Renaissance, these wars have unceasingly raised up national consciousness in Europe and educated generations of patriots'; the wars were 'the first European struggle for the independence of peoples and the liberty of men' (pp 83–4). In European history of all periods, according to Halecki (p 185), freedom is the basic and the key problem.

70 George Tatham, 'Environmentalism and Possibilism', in Griffith Taylor (ed), *Geography in the Twentieth Century* (London and New York 1957), p 155.

71 Isaiah Bowman, *Geography in Relation to the Social Sciences* (New York 1934), p 163.

72 Ibid, p 115.

73 Ibid, p 161.

74 Ibid, map p 38.

75 Curcio, p 5.

76 Ibid, p 7.

CHAPTER 3   GEOGRAPHY, CHINESE CIVILISATION,
AND WORLD HISTORY

1 Montesquieu, *De L'Esprit des Lois, Oeuvres Complètes*, 2 (Paris 1956–8), 5.14.

2 Ibid, 2.1.

3 Ibid, 12.10.

4 Ibid, 18.18.

5 Ibid, 5.14; cf 6.9, 8.8.

6 Ibid, 8.8; cf 8.2, 8.5, 8.10.

7 Ibid, 6.9. But here a footnote exempts China.

8 Ibid, 5.11.

9 Ibid, 5.14.

10 Ibid, 5.14, 6.1.

11 Ibid, 14.4, 19.12, 13.

12 Ibid, 19.14.

13 See above, p 26; also in Jean Bodin, *The Six Bookes of a*

Commonweale, Book 5 (Cambridge, Massachusetts, 1962; original 1576), p 562.

14 *De L'Esprit des Lois*, 14.2, 14.3.

15 Ibid, 16.8.

16 Ibid, 8.21, 23.13, 23.16. Religion also favours propagation, 23.21.

17 Ibid, 5.15, 17.2.

18 Ibid, 17.3.

19 Ibid, 17.3.

20 Ibid, 17.5; cf 18.19.

21 Ibid, 17.6, 8.19.

22 Ibid, 16.10.

23 Ibid, 12.29.

24 Ibid, 25.8.

25 Ibid, 18.6.

26 Ibid, 23.14, 8.21.

27 Ibid, 19.4.

28 Ibid, 17.8.

29 Ibid, 3.11.

30 Ibid, 11.6.

31 Ibid, 21.21.

32 Ibid, 20.21.

33 Ibid, 2.4. The nobility is the 'most natural' intermediate power between monarch and people (ibid).

34 *Oeuvres de Turgot*, 2 (Paris 1844), 597ff.

35 'Plan du second discours sur l'histoire universelle, dont l'objet sera les progrès de l'esprit humain', ibid, p 663.

36 'Plan du premier discours [sur l'histoire universelle], sur la formation des gouvernements et le mélange des nations', ibid, p 632.

37 Ibid, pp 632–3.

38 Georg Wilhelm Friedrich Hegel, *Vorlesungen über die Philosophie der Geschichte, Sämtliche Werke*, 11 (Stuttgart 1961), pp 120ff; Ernst Schulin, *Die weltgeschichtliche Erfassung des Orients bei Hegel und Ranke* (Göttingen 1958), pp 48–9; Martin Schwind, 'Die geographischen "Grundlagen" der Geschichte bei Herder, Hegel und Toynbee', *Erdkunde*, 14 (March 1960), 3–10.

39 Hegel, op cit, pp 134, 146–8.

40 Ibid, pp 255–6, 300

41 Ibid, p 147; Schulin, pp 45–6.

42 Hegel, pp 297–8.

43 Hegel, pp 145, 149–50, 159–60, 161; Schulin, pp 74–5. Single focus: eg, p 148.
44 Hegel, esp pp 296, 355; Schulin, pp 107–9.
45 Hegel, pp 159–60; Schulin, pp 74–5.
46 Cf Schulin, op cit, pp 52–3.
47 Hegel, p 120; Schulin, pp 58, 60, 134–5.
48 Though Schulin thinks Hegel is basically right; Schulin, 75 n 120, pp 136–7.
49 Hegel, p 191.
50 Otto Franke, *Geschichte des Chinesischen Reiches*, 1 (Berlin and Leipzig 1930), in his 'Vorwort', summarised the handling of China in German-language histories of the world.
51 Schulin, pp 141–2, 283–4.
52 Johann Gottfried von Herder, 'Ideen zur Philosophie der Geschichte der Menschheit' (1784), *Werke*, 3 (Leipzig nd), 343.
53 Ferdinand von Richthofen, *China*, 1 (Berlin 1877), 395–6.
54 George B. Cressey, *China's Geographic Foundations* (London and New York 1934), p 6.
55 Cressey, op cit, p 5, cites Franklin Thomas, *The Environmental Basis of Society* (New York and London 1925), p 7: '. . . the most important element in cultural progress is the contact of many cultures, while nothing breeds stagnation like isolation', etc.
56 'Classical religion' is C. K. Yang's term (*Religion in Chinese Society* (Berkeley 1961), p 23).
57 Geoffrey Barraclough, 'Universal History', in H. P. R. Finberg (ed), *Approaches to History* (London 1962), pp 84, 109.
58 Wolfgang Franke, *China und das Abendland* (Göttingen 1962), pp 122–3.
59 *Rome and China: A Study of Correlations in Historical Events* (Berkeley 1939), pp x–xi.
60 Marshall G. S. Hodgson, 'Hemispheric Interregional History as an Approach to World History', *Cahiers d'Histoire Mondiale*, 1 (January 1954), 715–23. Cf Barraclough, op cit, pp 102–6. Interest in trans-Pacific migrations provides a parallel in respect of the pre-Columbian culture of America.
61 Karl Jaspers, *Vom Ursprung und Ziel der Geschichte* (Zürich 1949), p 30.
62 Cited in Schulin, p 168.
63 Ibid, p 274.
64 This is the position of R. G. Collingwood; see especially his *Autobiography* (Oxford 1939), pp 114–15 and passim. Cf Henri-

Irénée Marrou, 'Comment comprendre le métier d'historien', in Charles Samaran (ed), *L'historie et ses méthodes* (Paris 1961), pp 1465–1540, esp p 1485.

65 E. M. Zhukov (ed), *Vsemirnaya Istoria v 10 tomakh*, 1 (Moscow 1955), xxi; by him also is E. Joukov, 'Des principes d'une histoire universelle', *Cahiers d'Histoire Mondiale*, 3 (1956), 527–35, covering much the same ground. Also I. S. Kon, *Die Geschichtsphilosophie des 20. Jahrhunderts: kritischer Abriss*, 2 (Berlin 1964), 221–86, 'Einheit und Vorwärtsschreiten der Weltgeschichte'; on this point, pp 222–3.

66 Zhukov, op cit, pp ix–x; Kon, op cit, pp 284–6; Wu T'ing-ch'iu, 'Establish a New System of World History', Peking *Kuang-ming Jih-pao*, 7–10 April 1961. O. V. Kuusinen et al, *Osnovy Marksizma-Leninizma* (Moscow 1960), pp 129–38.

67 See the polemic against Europocentrist history as an aspect of colonialism by Liu Ta-nien, 'Ya chou li shih tzen yang p'ing chia?', *Li shih yen chiu*, 3 (1965), 1–24. But even here Europe keeps its place in the sun in the period of early progressive capitalism, and also later when the European proletariat is most progressive.

68 Cf Marvin Harris, *The Rise of Anthropological Theory* (New York 1968), ch 22: 'Cultural Materialism: General Evolution'.

CHAPTER 4   WATERWORKS AND ASSOCIATED IDEAS

1 John King Fairbank, *The United States and China* (New York 1962), pp 47, 48–9.

2 C. W. Bishop, 'The Rise of Civilization in China with Reference to its Geographical Aspects', *Geographical Review*, 22 (1932), 619.

3 Cressey, op cit, pp 3–4.

4 Ibid, p 1. Similarly Fairbank, op cit, pp 23–7, 'The Harmony of Man and Nature'.

5 Cressey, op cit, p vii.

6 Ibid, p viii.

7 Ibid, p 3.

8 Arthur H. Smith, *Chinese Characteristics* (New York, Chicago and Toronto 1894), pp 320, 325, 330.

9 Karl Marx, *Das Kapital* (Berlin 1953), pp 538–9. See also the passage cited below, pp 77f. Wittfogel has a long and interesting discussion of the subject, with many quotations: 'Geopolitika, geograficheskiy materializm i marksizm', parts 1, 2 and 3, *Pod*

znamenem marksizma, 2–3 (February–March 1929), 16–42; 6 (June), 1–29; 7–8 (July–August), 1–28. See also Ian M. Matley, 'The Marxist Approach to the Geographical Environment', *Annals of the Association of American Geographers*, 56 (March 1966), 97–111, and the ensuing discussion in vol 57 of the *Annals* (March 1967), 203–7.

10   F. Engels, 'The Part Played by Labour . . .', in K. Marx and F. Engels, *Selected Works* (Moscow 1968), pp 365–6.

11   Cited in Gustav A. Wetter, *Dialectical Materialism* (New York 1958), p 107.

12   K. A. Wittfogel, 'Die natürlichen Ursachen der Wirtschaftsgeschichte', *Archiv für Sozialwissenschaft und Sozialpolitik*, 67 (1932), 731.

13   J. Stalin, 'Dialectical and Historical Materialism', in his *Problems of Leninism* (Moscow 1945), p 582. Cf Matley, op cit, pp 101–2; I. I. Ivanov-Omskiy, *Istoricheskiy materializm o roli geograficheskoy sredy v razvitii obshchestva* (Moscow 1950), pp 9–10, 14.

14   The debate can be followed in *Soviet Geography: Review and Translation* starting with Yu. G. Saushkin's article 'The geographical environment of human society' in vol 4 (December 1963), 3–19.

15   'Marx . . . distinguishes two aspects of the mode of production: the relation of man to Nature is determined by his *productive forces*; the mutual relationships of men in the process of production find expression in the *relations of production*.' Gustav A. Wetter, *Soviet Ideology Today* (London 1966), pp 161–2. 'These two categories,' Wetter adds, 'represent the fundamental concepts of historical materialism' (p 162). See also O. V. Kuusinen et al, *Osnovy Marksizma-Leninizma* (Moscow 1960), pp 121–3.

16   Stalin, op cit, pp 583–95.

17   Karl Marx, 'The British Rule in India', in Karl Marx and Friedrich Engels, *Basic Writings on Politics and Philosophy*, ed Lewis S. Feuer (Garden City, New York 1959), pp 477–81.

18   George Lichtheim, *The Concept of Ideology and Other Essays* (New York 1967), 'Oriental despotism', pp 62–93 (originally published as 'Marx and the "Asiatic Mode of Production" ', *St Anthony's Papers*, 14 (Oxford and London 1963); on this point, 88–9.

19   *Chung kuo kung ch'an tang ti liu tz'u ch'üan kuo ta hui i chüeh an* (1928), pp 150–1.

20   On these various questions, see Obshchestvo Marksistov-

vostokovedov, *Diskussiya ob aziatskom sposobe proizvodstva; po dokladu M. Godesa* (Moscow and Leningrad 1931), passim, esp pp 8, 15, 18, 34, 35, 38, 50, 52–4, 60, 66–7, 110. Also Wittfogel, *Oriental Despotism: A Comparative Study of Total Power* (New Haven 1959), pp 402ff.

21 Cf Stuart Schram and Hélène Carrère d'Encausse: *Le Marxisme et l'Asie, 1853–1964* (Paris 1965), pp 130–2. The revival can be most conveniently followed in *La Pensée* and *Narody Azii i Afriki* from the mid-1960s on.

22 Marx, 'The British Rule in India', op cit, p 476. Cf *Das Kapital*, p 539.

23 Eg, recently by E. Varga, *Ocherki po problemam politekonomii kapitalizma* (Moscow 1965), p 361. According to V. A. Rubin, 'Problemy vostochnoy despotii v rabotakh sovetskikh issledovateley', *Narody Azii i Afriki*, 4 (1966), 98–9, several Soviet authors in recent years have argued the interdependence of climate, irrigation, bureaucracy, and despotism.

24 Cf Wittfogel, *Oriental Despotism*, introduction and ch 9.

25 *Influences of Geographic Environment* (New York 1911), p 328.

26 Wittfogel, *Wirtschaft und Gesellschaft Chinas; Versuch der wissenschaftlichen Analyse einer grossen asiatischen Agrargesellschaft; 1. Teil: Produktivkräfte, Produktions- und Zirkulationsprozess* (Leipzig, 1931), p 187.

27 Wittfogel, *Das erwachende China; ein Abriss der Geschichte und der gegenwärtigen Probleme Chinas* (Vienna 1926), p 16.

28 Wittfogel, *Wirtschaft und Gesellschaft*, pp 125–6.

29 Ibid, pp 278–9.

30 Ibid, pp 284–5.

31 Ibid, pp 425.

32 Wittfogel, 'The Foundations and Stages of Chinese Economic History', *Zeitschrift für Sozialforschung*, 4 (1935), 41.

33 Ibid, 46.

34 Wittfogel, *New Light on Chinese Society: An Investigation of China's Socio-Economic Structure* (New York 1938), p 17; idem, 'Die Theorie der orientalischen Gesellschaft', *Zeitschrift für Sozialforschung*, 7 (1938), 108.

35 Wittfogel, 'Foundations and Stages', loc cit, p 50.

36 Wittfogel, *Oriental Despotism*, p 33.

37 Wittfogel, 'Imperial China—A "Complex" Hydraulic (Oriental) Society', in John Meskill (ed), *The Pattern of Chinese History* (Boston 1965), p 87.

38 Wittfogel, *Wirtschaft and Gesellschaft*, pp 425–6.
39 Wittfogel, 'Foundations and Stages', op cit, pp 51, 52.
40 Wittfogel, 'Imperial China', op cit, p 86; see also idem, 'Ideas and the Power Structure', in Wm Theodore de Bary and Ainslee T. Embree (eds), *Approaches to Asian Civilizations* (New York 1964), p 95.
41 Wittfogel, *Oriental Despotism*, pp 169–70 (italics added). Cf 'Theorie', 108–9.
42 Wittfogel, 'Theorie', loc cit, pp 99–100; *Oriental Despotism*, pp 15–19.
43 Wittfogel, 'Ideas', p 87.
44 Wittfogel, 'Results and Problems of the Study of Oriental Despotism', *Journal of Asian Studies*, 28 (February 1969), 361.
45 Wittfogel, 'Imperial China', op cit, p 92.
46 Wittfogel, *Oriental Despotism*, p 21.
47 Wittfogel, *Wirtschaft und Gesellschaft*, pp 280, 282.
48 Shih Nien-hai, *Ho shan chi* (Peking 1963), pp 131, 138, 145ff, and maps pp 88, 164; Tsou Pao-chün, *Ti hsüeh t'ung lun* (Taipei 1965), pp 406–7; maps of Shang and Chou sites in Chêng Tê-k'un, *Archaeology in China*, 2: *Shang China* (Cambridge 1960), p 14, and 3: *Chou China* (Cambridge 1963), p 8; maps in Chang Kwang-chih, *The Archaeology of Ancient China* (New Haven 1968), pp 195, 271, 277, 278; Sun Ching-chih et al, *Severnyi Kitay* (Moscow 1958), pp 72–3, 224.
49 Henri Maspero, 'Contribution à l'étude de la société chinoise à la fin des Chang et au début des Tcheou', *Bulletin de l'École Française d'Extrême-Orient*, 45, fasc 2 (1954), 335–403; Hermann von Wissmann, 'On the Role of Nature and Man in Changing the Face of the Dry Belt of Asia', in William L. Thomas, Jr (ed), *Man's Role in Changing the Face of the Earth* (Chicago 1956), pp 278–303, esp the map on p 289; Shih Nien-hai, op cit, pp 139–40, 145–6; T'ien chin shih wen hua chü k'ao ku fa chüeh tui, 'Po hai wan hsi an k'ao ku tiao ch'a ho hai an hsien pien ch'ien yen chiu', *Li shih yen chiu*, 1 (1966), 58, 62; *Chiu ho: Tz'u hai*, 1, 107c–d; Chang Kwang-chih, *The Archaeology of Ancient China* (New Haven 1968), pp 34–5.
50 Shen Ping, *Huang ho t'ung k'ao* (Taipei 1960), p 16.
51 Ibid, p 17.
52 Ibid, p 43.
53 Ibid, p 44; Chang Han-ying, *Li tai chih Ho fang lüeh shu yao* (Chungking 1945), p 8.

54 Sung Hsi-shang et al, *Chung kuo ho ch'uan chih*, 1 (Taipei 1955), 37; Cheng Chao-ching, *Chung kuo shui li shih* (Taipei 1966), p 10.

55 Shih Nien-hai, op cit, pp 97–101. The basic source of information on these early works is Szu-ma Ch'ien's *Shih chi*, ch 29.

56 Wittfogel, *Wirtschaft und Gesellschaft*, p 415.

57 Wittfogel, 'Ideas', op cit, p 95; cf idem, *Oriental Despotism*, p 24.

58 Wittfogel, *Oriental Despotism*, p 15ff.

59 Ibid, p 33.

60 Wittfogel, 'Imperial China', op cit, p 87; see also idem, *Wirtschaft und Gesellschaft*, pp 83–4, 205–6, 228–9.

61 Glenn T. Trewartha, 'Ratio Maps of China's Farms and Crops', *Geographical Review*, 28 (1938), 109; Sun Ching-chih et al, op cit, map p 63.

62 David Noel Keightley, *Public Work in Ancient China: A Study of Forced Labor in the Shang and Western Chou* (unpublished dissertation, New York 1969), pp 124, 313. Cf Chêng Tê-k'un, *Shang China*, p 197: 'There is nothing to indicate the possible use of an irrigation system, not even a water-well. The crops depended entirely on the weather'; but Chang Kwang-chih, op cit, p 249: 'irrigation was probably employed'.

63 Sun Ching-chih et al, op cit, p 73; Dwight H. Perkins, *Agricultural Development in China, 1368–1968* (Chicago 1969), pp 62–3.

64 China International Famine Relief Commission, *Well Irrigation in West Hopei; Preliminary Report*, Series B, 49 (Peiping 1931), 4. See also Wolfram Eberhard, *Conquerors and Rulers; Social Forces in Medieval China* (Leiden 1952), pp 34–43; Shen Tsung-han, *Chung kuo nung yeh tzu yüan*, 1 (Taipei 1952), 104–11; Sidney D. Gamble, *Ting Hsien; A North China Rural Community* (New York 1954), pp 230–6; John Lossing Buck (ed), *Land Utilization in China* (New York 1964), p 188; Perkins, op cit, p 68; Pierre Gourou, letter in *Annals of the Association of American Geographers*, 51 (December 1961), 401–2; Sun Ching-chih et al, op cit, p 64.

65 Wittfogel, *Wirtschaft und Gesellschaft*, pp 189–300, 454.

66 Ibid, pp 187, 189.

67 Ibid, p 251.

68 Wittfogel, 'Theorie', op cit, p 99; 'Foundations and Stages', op cit, p 58; *Wirtschaft und Gesellschaft*, p 130.

69 Wittfogel, 'Theorie', op cit, p 114.

70 Wittfogel, *Wirtschaft und Gesellschaft*, pp 64–5.

71 Ibid, p 30. The broad pattern of reasoning is visible in the overall organisation of subjects in this first volume and the pro-

jected (but never published) second one; the sequence is from the environment—geology, climate and so forth, and their causes—through to classes, state, and ideology (ibid, pp xiii–xiv). In principle, Wittfogel allows for retroaction, in accordance with Marx (eg, ibid, p 2), but in application he makes the one-way action decisive.

72 Wittfogel and Fêng Chia-shêng: *History of Chinese Society: Liao (907–1125)* (Philadelphia 1949), p 41. Cf Wittfogel, *Wirtschaft und Gesellschaft*, pp 1–2.

73 This is worked out most faithfully in 'Geopolitika'. In *Oriental Despotism* he is much less concerned with geography, and while keeping the man/nature distinction as part of his thinking, speaks of 'the primary importance of institutional (and cultural) factors' (p 11), and opportunity, not necessity (p 12). But more often his style remains deterministic: under broad institutional conditions, 'man, reacting specifically to the water-deficient landscape, moves toward a specific hydraulic order of life' (p 12) and under these conditions 'this natural configuration decisively affected man's behaviour as a provider of food and organizer of human relations' (p 13).

74 Wittfogel, 'Geopolitika', Part 2, pp 19ff.

75 See above, p 71.

76 Wittfogel, *Oriental Despotism*, p 161.

77 Ibid, p 441.

78 Ibid, p 448.

79 Op cit, pp 4–5.

80 Wittfogel, *Oriental Despotism*, p 17.

81 See especially his *Inner Asian Frontiers of China* (Boston 1962); and *Studies in Frontier History: Collected Papers, 1928–1958* (London 1962).

82 Lattimore, *Inner Asian Frontiers*, p 340.

83 Lattimore, *Studies*, pp 494–5.

84 Cf Stuart Schram and H. Carrère d'Encausse, *Le Marxisme et l'Asie, 1853–1964* (Paris 1965), p 131. Jean Chesneaux's denial of the anti-Chinese import of the revival of Marxist discussion of the Asiatic mode ('Où en est la discussion sur le "mode de production asiatique"?', *La Pensée*, 129 (July–August 1966), 39–40) misses the point that while the revival may not have an immediate political cause, it can hardly help having this political function.

CHAPTER 5   NATURAL RESOURCES

1 Erich W. Zimmermann, *Introduction to World Resources* (New York 1964).
2 Jean Baptiste du Halde, *Description Géographique, Historique, Chronologique, Politique, et Physique de l'Empire de la Chine et de la Tartarie Chinoise* (Paris 1735).
3 Jedidiah Morse, *Geography Made Easy* (Boston 1806), pp 375, 379.
4 Edwin Joshua Dukes, *Everyday Life in China* (London 1885), p 158. Cf F. von Richthofen, *Baron Richthofen's Letters, 1870–1872* (Shanghai 1903).
5 du Halde, op cit, 2, 145.
6 François Quesnay, *Despotism in China* (1767), in Lewis A. Maverick, *China, A Model for Europe* (San Antonio 1946), 1.3, 7.1, 7.5.
7 T. R. Malthus, *An Essay on the Principle of Population*, 2 (London 1806), 203, 205.
8 Cressey, *China's Geographic Foundations*, p 23. A detailed criticism of Cressey's thought in this and other respects is Hsü Chao-k'uei, 'Ko Te-shih fan tung ti li hsüeh szu hsiang p'i p'an', *Acta Geographica Sinica*, 24 (February 1958), 103–17.
9 Cressey, op cit, p 110.
10 Cressey, *Land of the 500 Million* (New York 1955), pp 130, 347.
11 Central Intelligence Agency, *Communist China Map Folio* (1967), texts to maps of 'Fuels and Power' and 'Minerals and Metals'. See also K. P. Wang, 'Rich Minerals Resources Spur Communist China's Bid for Industrial Power', *Mineral Trade Notes, Special Supplement*, 59 (March 1960); idem, 'The Mineral Resource Base of Communist China', in Joint Economic Committee, Congress of the United States, *An Economic Profile of Mainland China*, 1 (Washington 1967), 170, table 1; and 193.
12 Cressey, op cit, p 118; idem, *Land of the 500 Million*, pp 141–2. Wang, 'Rich Mineral Resources', loc cit, p 23.
13 Cressey, *Land of the 500 Million*, p 106; Marion R. Larsen, 'China's Agriculture Under Communism', in Joint Economic Committee, Congress of the United States, *An Economic Profile of Mainland China*, 1, 207–8; Chu Chien-nung, *T'u ti fei li ching chi yüan li* (Shanghai 1964), p 144. Ho Ping-ti, *Studies on the Population of China, 1368–1953* (Cambridge, Mass 1959), pp 132, 135, thinks 267 million is probably too low a figure.

14 Cressey, *Land of the 500 Million*, pp 101–2.
15 G. Etzel Pearcy, 'Mainland China—Geographic Strengths and Weaknesses', *Department of State Bulletin*, 55 (29 August 1966), 298.
16 John S. Aird, 'Population Growth and Distribution in Mainland China', in Joint Economic Committee, Congress of the United States, *An Economic Profile of Mainland China*, 2, 373.
17 Famine in India in 1966 and 1967 struck particularly in Orissa, Bihar, Uttar Pradesh, Madhya Pradesh and Andhra Pradesh, to judge by the current accounts in the *New York Times*; the death of millions was averted only by massive emergency grain imports. In China, there have been occasional food shortages since 1949 but no famine and, unlike the United States, no starvation; see Felix Greene, *A Curtain of Ignorance* (London 1968), ch 6, 'The Starving Chinese'.
18 Irene B. Taeuber, 'China's Population: Riddle of the Past, Enigma of the Future', in Albert Feuerwerker (ed), *Modern China* (Englewood Cliffs 1964), p 22.
19 Eadem, 'Asia Populations: The Critical Decades', in Larry K. Y. Ng (ed), *The Population Crisis* (Bloomington 1965), p 87.
20 Warren S. Thompson, *Population and Progress in the Far East* (Chicago 1959), pp 270–1.
21 Ibid, p 402. Cf Greene, op cit, esp ch 9.
22 Preston E. James, *One World Divided: A Geographer Looks at the Modern World* (New York 1964), p 343.
23 Albert H. Rose, *A Geography of International Relations* (1965), p 385.
24 Taeuber, 'Asian Populations', op cit, pp 86–7; Thompson, op cit, pp 391–6. Cf Rhoads Murphey, 'China and the Dominoes', *Asian Survey*, 6 (September 1966), 510–15.
25 Eg, Wang Chun-heng, *A Simple Geography of China* (Peking 1958); Jen Yu-ti, *A Concise Geography of China* (Peking 1964); Chu Chien-nung, op cit, p 144; Mao Tse-tung, 'On Protracted War', *Selected Works*, 2 (Peking 1967), 123.
26 Mao Tse-tung, 'On Contradiction', *Selected Works*, 1 (Peking 1965), 312, 314.
27 Wang Chun-heng, op cit, pp 55–7.
28 Li Lin-ku, 'She hui sheng ch'an fang shih ho jen k'ou wen t'i', *Hsin chien she*, 4 (1960), 51.
29 US Government, Department of State, *United States Relations with China* (Washington 1949), pp iv–v.

30 Mao Tse-tung, 'The Bankruptcy of the Idealist Conception of History', *Selected Works*, 4 (Peking 1967), 453.

31 Ibid, p 454. Cf Chu Chien-nung, op cit, pp 220–32.

32 Chu Chien-nung, op cit, pp 134, 223, 228, 242.

33 Mao Tse-tung, 'On the Correct Handling of Contradictions Among the People', in *Selected Readings from the Works of Mao Tse-tung* (Peking 1967), p 373. The speech was given on 27 February 1957.

34 *New York Times*, 13 June 1957, reporting summaries of Mao's speeches of 27 February and 12 March 1957.

35 Edgar Snow, 'Population Control in China: An Interview with Chou En-lai', in Ng, op cit, p 102. Leo Orleans (Library of Congress) estimates a present rate of increase of 1·5 or 1·6 per cent (*New York Times*, 25 January 1970). In January 1973 the *New York Times* reported that a new birth-control drive was under way (28 January 1973).

36 Mao Tse-tung, 'On Protracted War', op cit, pp 143, 151.

37 *Hsüeh-hsi Mao Tse-tung ti szu-hsiang fang-fa ho kung-tso fang-fa* (Peking 1958), p 73, cited in Stuart Schram, *Mao Tse-tung* (Baltimore nd), p 295.

38 'Reply to Comrade Kuo Mo-jo', *Ten More Poems of Mao Tse-tung* (Hong Kong 1967), p 14. The novel, based on old tradition, was written by Wu Ch'eng-en in the sixteenth century. Arthur Waley translated it into English as *Monkey* (New York 1958).

39 'Natural Resources and Economic Development', *Annals of the Association of American Geographers*, 47 (September 1957), 212. Cf H. B. Chenery, 'The Effects of Resources on Economic Growth', in Kenneth Berrill (ed), *Economic Development with Special Reference to East Asia* (London 1965), esp pp 46–9.

40 Harold J. Barnett and Chandler Morse, *Scarcity and Growth: The Economics of Natural Resource Availability* (Baltimore 1963), p 11.

41 Gunnar Myrdal, *Asian Drama*, 3 (New York 1968), 2066–8.

42 'Main Lines of Chinese Communist Economic Policy', in Joint Economic Committee, Congress of the United States, *An Economic Profile of Mainland China*, 1, 27–8.

43 President's Science Advisory Committee, *The World Food Problem*, 2 (Washington 1967), 195; Kenneth E. F. Watt, *Ecology and Resource Management; A Quantitative Approach* (New York 1968), pp 12ff.

44 Barnett and Morse, op cit, pp 235–6. Cf Engels's observation in *Outlines of a Critique of Political Economy* (1844) that the progress

of science is 'just as limitless and at least as rapid as that of population', Ronald L. Meek, *Marx and Engels on Malthus* (New York 1954), p 63.

CHAPTER 6 CONCLUSION

1 *Newsweek* (21 February 1972).
2 *New York Review of Books* (24 February 1972).
3 Ibid.

BIBLIOGRAPHY

Aird, John S. 'Population Growth and Distribution in Mainland China', in Joint Economic Committee, Congress of the United States, *An Economic Profile of Mainland China*, 2 (Washington 1967), 341–401

Aristotle. *Politics*, trans Benjamin Jowett (New York 1943)

Ashbrook, Arthur G., Jr. 'Main Lines of Chinese Communist Economic Policy', in Joint Economic Committee, Congress of the United States, *An Economic Profile of Mainland China*, 1 (Washington 1967), 15–44

Augustine. *De Civitate Dei* (Migne, 41).

Bacon, Roger. *Opus Majus*, ed John Henry Bridges, 2 vols (Oxford 1897)

Bacon, Roger. *The* Opus Majus *of Roger Bacon*, trans Robert Belle Burke, 2 vols (1928, reprinted New York 1962)

Balazs, Étienne. *Chinese Civilization and Bureaucracy*, ed Arthur F. Wright, trans H. M. Wright (New Haven 1964)

Barker, Ernest, et al (eds). *The European Inheritance*, 3 vols (Oxford 1954)

Barnett, Harold J., and Morse, Chandler. *Scarcity and Growth; the Economics of Natural Resource Availability* (Baltimore 1963)

Barraclough, G. 'Universal History', in H. P. R. Finberg (ed), *Approaches to History* (London 1962), 83–109

Beazley, C. Raymond. *The Dawn of Modern Geography*, 1–3 (1897–1906, reprinted New York 1949)

Bishop, C. W. 'The Rise of Civilization in China with Reference to its Geographical Aspects', *Geographical Review*, 22 (1932), 617–31

Black, C. E. *The Dynamics of Modernization: A Study in Comparative History* (New York 1966, reprinted 1967)

Bodde, Derk. *China's First Unifier* (Leiden 1938)

Bodin, Jean. *The Six Bookes of a Commonweale*, a Facsimile Reprint of the English Translation of 1606, trans Richard Knolles (Cambridge 1962)

Bowman, Isaiah. *Geography in Relation to the Social Sciences*, American Historical Association, Report of the Commission on the Social Studies, Part 5 (New York 1934)

Braudel, Fernand. 'La géographie face aux sciences humaines', *Annales-Économies, Sociétés, Civilisations*, 6 (1951), 485–92

Buchanan, Keith. *The Chinese People and the Chinese Earth* (London 1966)

Buck, John Lossing (ed). *Land Utilization in China* (1937, reprinted New York 1964)

Bunbury, E. H. *A History of Ancient Geography*, 2nd ed (London 1883)

Bunge, William. *Theoretical Geography*, Lund Studies in Geography, Series C, General and Mathematical Geography, no 1, Department of Geography, Royal University of Lund, 2nd ed (Lund 1966)

Central Intelligence Agency, US Government, *Communist China Map Folio* (1967)

Chang Kwang-chih, *The Archaeology of Ancient China*, 2nd ed (New Haven 1968)

Chenery, H. B. 'The Effects of Resources on Economic Growth', in Kenneth Berrill (ed), *Economic Development with Special Reference to East Asia* (London 1965), 19–52

Chêng Tê-k'un. *Archaeology in China*, 2: *Shang China*; 3: *Chou China* (Cambridge 1960, 1963)

Chesneaux, Jean. 'Où en est la discussion sur le "mode de production asiatique"?', *La Pensée*, no 122 (August 1965), 40–59; no 129 (July–August 1966), 33–46

Chi Ch'ao-ting. *Key Economic Areas in Chinese History, as Revealed in the Development of Public Works for Water-control* (London 1936)

China International Famine Relief Commission. *Well Irrigation in West Hopei; Preliminary Report*, Series B, no 49 (Peiping 1931)

Chu Shih-chia. *Chang Hsüeh-ch'êng, his Contributions to Chinese Local Historiography*, unpublished PhD dissertation, Columbia University (1950)

Coedès, George. *Textes d'auteurs grecs et latins relatifs à l'Extrême-Orient, depuis le IVe siècle av. J.-C., jusqu'au XIVe siècle* (Paris 1910)

Collingwood, R. G. *An Autobiography* (Oxford 1939)

Cressey, George Babcock. *China's Geographic Foundations; A Survey of the Land and its People* (New York and London 1934)

Cressey, George B. *Asia's Lands and Peoples*, 2nd ed (New York 1951)

Cressey, George B. *Land of the 500 Million* (New York 1955)

Curcio, Carlo. *Europa: Storia di un'idea*, 2 vols (Florence 1958)

Dawson, Christopher. *The Making of Europe: An Introduction to the History of European Unity* (1932, reprinted London 1939)

Dawson, Raymond. *The Chinese Chameleon: An Analysis of European Conceptions of Chinese Civilization* (New York 1967)

De Bary, W. T. (ed). *Sources of Chinese Tradition* (New York 1960)

Demiéville, P. 'Chang Hsüeh-ch'eng and his Historiography', in W. G. Beasley and E. G. Pulleyblank (eds), *Historians of China and Japan* (London 1961), 167–85

Demolins, Edmond. *Comment la route crée le type social*, 2 vols (Paris nd (1901?))

Diez del Corral, Luis. *El rapto de Europa: una interpretación historica de nuestro tiempo* (Madrid 1954)

Du Halde, Jean Baptiste. *Description géographique, historique, chronologique, politique, et physique de l'empire de la Chine et de la Tartarie chinoise*, 4 vols (Paris 1735)

Dukes, Edwin Joshua. *Everyday Life in China; or Scenes Along River and Road in Fuh-kien* (London 1885)

Eberhard, Wolfram. *Conquerors and Rulers; Social Forces in Medieval China* (Leiden 1952)

Erikson, Erik H. *Childhood and Society*, 2nd ed (New York 1963)

Fairbank, John King (ed). *Chinese Thought and Institutions* (Chicago 1957)

Fairbank, John King. *The United States and China*, 2nd ed (New York 1962)

Fischer, Bonifatius. *Vetus Latina: die Reste der altLateinischen Bibel; 2, Genesis* (Freiburg 1954)

Fischer, Jürgen. *Oriens—Occidens—Europa: Begriff und Gedanke 'Europa' in der späten Antike und im frühen Mittelalter* [5th to 11th centuries], Veröffentlichungen des Instituts für europäische Geschichte Mainz, Band 15, Abteiling Universalgeschichte, ed Martin Göhring (Wiesbaden 1957)

FitzGerald, C. P. *China: A Short Cultural History*, 3rd ed (London 1965)

Franke, Otto. *Geschichte des Chinesischen Reiches*, 1 (Berlin and Leipzig 1930)

Franke, Wolfgang. *China und das Abendland* (Göttingen 1962)

Freising, Otto Bischof von. *Chronik oder die Geschichte der zwei Staaten,* trans Adolf Schmidt, ed Walther Lammers, Latin and German texts (Berlin 1960)

Fried, Morton H. *The Evolution of Political Society: An Essay in Political Anthropology* (New York 1967)

Fritzemeyer, Werner. *Christenheit und Europa; zur Geschichte des europäischen Gemeinschaftsgefühls von Dante bis Leibniz*, Beiheft 23 der *Historischen Zeitschrift* (1931)

Fung Yu-lan. *A Short History of Chinese Philosophy* (New York 1948)

Fung Yu-lan. *A History of Chinese Philosophy*, trans Derk Bodde, 2 vols (Princeton 1952 (1), 1953 (2))

Gamble, Sidney D. *Ting Hsien; A North China Rural Community* (New York 1954)

Gerald of Wales. *The Historical Works of Giraldus [de Barri] Cambrensis, containing the Topography of Ireland, &c.*, trans Thomas Forester, revised and ed Thomas Wright (London 1863). *See also* Giraldus Cambrensis

Ginsburg, Norton. 'Natural Resources and Economic Development', *Annals of the Association of American Geographers*, 47 (September 1957), 197–212

Giraldus Cambrensis. *Topographia hibernica, et Expugnatio hibernica*, ed James F. Dimock (London 1867), vol 5 of *Giraldi Cambrensis Opera*, no 21 in *Rerum britannicarum medii aevi scriptores*, or Chronicles and Memorials of Great Britain and Ireland during the Middle Ages, published by the Authority of Her Majesty's Treasury under the Direction of the Master of the Rolls

Gollwitzer, Heinz. *Europabild und Europagedanke: Beiträge zur deutschen Geistesgeschichte des 18. und 19. Jahrhunderts*, 2nd ed, revised (Munich 1964)

Gould, Peter R. 'Man Against His Environment: A Game Theoretic Framework', *Annals of the Association of American Geographers*, 53 (1963), 290–7

Granet, Marcel. *La pensée chinoise* (1934, reprinted Paris 1950)

Gray, J. 'Historical Writing in Twentieth-century China: Notes on its Background and Development', in W. G. Beasley and E. G. Pulleyblank (eds), *Historians of China and Japan* (London 1961), 186–212

Green, Felix. *A Curtain of Ignorance* (London 1968)

Halecki, Oscar. *The Limits and Divisions of European History* (New York 1950)

Harris, Marvin. *The Rise of Anthropological Theory* (New York 1968)

Hartshorne, Richard. *The Nature of Geography* (reprinted from

Annals of the Association of American Geographers, 29 (1939)) (Lancaster, Pennsylvania 1961)

Hay, Denys. *Europe: The Emergence of an Idea* (Edinburgh 1957)

Hegel, Georg Wilhelm Friedrich. *Sämtliche Werke*, ed Hermann Glockner, 11, *Vorlesungen über die Philosophie der Geschichte* (Stuttgart 1961)

Herder, Johann Gottfried von. *Herders Werke*, ed H. Kurz (Leipzig nd (1850?))

Herodotus. *The Persian Wars*, trans George Rawlinson (New York 1942)

Higgins, Benjamin. *Economic Development: Principles, Problems, and Policies* (New York 1968)

Hippocrates. *Airs Waters Places*, trans W. H. S. Jones (London 1923)

Ho Ping-ti. *Studies on the Population of China, 1368–1953* (Cambridge 1959)

Hodgson, Marshall G. S. 'Hemispheric Interregional History as an Approach to World History', *Cahiers d'Histoire Mondiale*, 1 (January 1954), 715–23.

Isaacs, Harold R. *Images of Asia: American Views of China and India*, 2nd ed (New York 1962)

Isidore of Seville. *Etymologiae* (Migne, 82)

Ivanov-Omskiy, I. I. *Istoricheskiy materializm o roli geograficheskoy sredy v razvitii obshchestva* (Moscow 1950)

James, Preston E. *One World Divided; A Geographer Looks at the Modern World* (New York 1964)

Jaspers, Karl. *Vom Ursprung und Ziel der Geschichte* (Zürich 1949)

Jen Yu-ti. *A Concise Geography of China* (Peking 1964)

Joint Economic Committee, Congress of the United States. *An Economic Profile of Mainland China; Studies Prepared for the Joint Economic Committee*, 2 vols (Washington 1967)

Joukov, E., *see* Zhukov, E. M.

Juillard, Étienne. 'Aux frontières de l'histoire et de la géographie', *Revue historique*, 215 (1956), 267–73

Karlgren, Bernhard. 'The Book of Documents', Ostasiatiska Samlingarna (Museum of Far Eastern Antiquities, Stockholm), *Bulletin*, 22 (1950), 1–81

Karlgren, Bernhard. 'Grammata Serica Recensa', Ostasiatiska Samlingarna (Museum of Far Eastern Antiquities, Stockholm), *Bulletin*, 29 (1957)

Keightley, David Noel. *Public Work in Ancient China: A Study of*

Forced Labor in the Shang and Western Chou, unpublished PhD dissertation, Columbia University (New York 1969)

Kon, I. S. *Die Geschichtsphilosophie des 20. Jahrhunderts; Kritischer Abriss*, trans from Russian by W. Hoepp, 2 vols (Berlin 1964)

Kuusinen, O. V., et al. *Osnovy Marksizma-Leninizma* (Moscow 1960)

Lach, Donald F. *Asia in the Making of Europe*, 1: *The Century of Discovery*, books 1 and 2 (Chicago and London 1965)

Larsen, Marion R. 'China's Agriculture Under Communism', in Joint Economic Committee, Congress of the United States, *An Economic Profile of Mainland China*, 1 (Washington 1967), 197–267

Lattimore, Owen. *Inner Asian Frontiers of China* (1940, reprinted Boston 1962)

Lattimore, Owen. *Studies in Frontier History; Collected Papers, 1928–1958* (London 1962)

Legge, James. *The Chinese Classics*, 1 (Oxford: Clarendon Press, 2nd ed, 1893); 2 (Hong Kong and London: Trübner, 1861)

Legge, James. *The Sacred Books of China; The Texts of Confucianism*, parts 3 and 4, the *Lî Kî*, 2 vols (Oxford 1885)

Levenson, Joseph R. *Confucian China and its Modern Fate: A Trilogy*, 3 vols in 1. Vol 1: *The Problem of Intellectual Continuity*; 2: *The Problem of Monarchical Decay*; 3: *The Problem of Historical Significance* (Berkeley, California, 1968)

Lichtheim, George. *The Concept of Ideology, and Other Essays* (New York 1967)

MacIver, R. M. *Social Causation* (1942, reprinted New York 1964)

Malthus, T. R. *An Essay on the Principle of Population . . .*, 2 vols, 3rd ed (London 1806)

Mao Tse-tung. *Selected Readings from the Works of Mao Tse-tung* (Peking 1967)

Mao Tse-tung. *Selected Works of Mao Tse-tung* (Peking 1965 (1), 1967 (2, 3 and 4))

Mao Tse-tung. *Ten More Poems of Mao Tse-tung* (Hong Kong 1967)

Marrou, Henri-Irénée. 'Comment comprendre le métier d'historien', in Charles Samaran (ed), *L'histoire et ses méthodes* (Paris 1961), pp 1465–1540

Marx, Karl. 'The British Rule in India' (1853), in Karl Marx and Friedrich Engels, *Basic Writings on Politics and Philosophy*, ed Lewis S. Feuer (New York 1959), 474–81

Marx, Karl. *Das Kapital; Kritik der politischen Ökonomie*, 3 vols (Berlin 1953)

Marx, Karl, and Engels, Friedrich. *Selected Works* (Moscow 1968)

Maspero, Henri. 'Contribution à l'étude de la société chinoise à la fin des Chang et au début des Tcheou', *Bulletin de l'École Française d'Extrême-Orient*, 46 (1954), Fasc 2, 335–403

Matley, Ian M. 'The Marxist Approach to the Geographical Environment', *Annals of the Association of American Geographers*, 56 (March 1966), 97–111

Maverick, Lewis A. *China, A Model for Europe* (San Antonio 1946)

Meek, Ronald L. *Marx and Engels on Malthus* (New York 1954)

Migne: Jacques Paul Migne (ed). *Patrologiae cursus completus . . . series latina* (Paris 1844–80)

Mill, John Stuart. *On Liberty* (1859, reprinted Chicago 1955)

Minshull, Roger. *Regional Geography: Theory and Practice* (London 1967)

Montesquieu, Charles-Louis. *Oeuvres Complètes*, 2 vols, Pléiade ed (Paris 1956–8)

Morse, Jedidiah. *Geography Made Easy; Being an Abridgment of the American Universal Geography*, 10th ed (Boston 1806)

Murphey, Rhoads. 'China and the Dominoes', *Asian Survey*, 6 (September 1966), 510–15

Murphey, Rhoads. 'Man and Nature in China', *Modern Asian Studies*, 1 (1967), 313–33

Myrdal, Gunnar. *Asian Drama; an Inquiry into the Poverty of Nations*, 3 vols (New York 1968)

Needham, Joseph. *Science and Civilisation in China*, to comprise seven vols (Cambridge: 1 (1954), 2 (1956), 3 (1959), 4 Part 1 (1962), 4 Part 2 (1965))

Ng, Larry K. Y. (ed). *The Population Crisis: Implications and Plans for Action* (Bloomington, Indiana 1965)

Obshchestvo Marksistov-Vostokovedov. *Diskussiya ob aziatskom sposobe proizvodstva; po dokladu M. Godesa* (Moscow 1931)

Orosius. *Historiae adversus paganos*, in *Corpus scriptorum ecclesiasticorum latinorum*, 5 (Vienna 1882)

Paassen, C. van. *The Classical Tradition of Geography* (Groningen 1957)

Parkinson, C. Northcote. *East and West* (1963, reprinted New York 1965)

Paulys Realencyclopädie der classischen Altertums-wissenschaft, ed Georg Wissowa, 2 (Stuttgart 1896, sv 'Asia')

Pearcy, G. Etzel. 'Mainland China—Geographic Strengths and Weaknesses', *Department of State Bulletin*, 55 (29 August 1966), 294–303

Perkins, Dwight H. *Agricultural Development in China, 1368–1968* (Chicago 1969)

Piccolomini, Aeneas Sylvius. *Enea Silvio Piccolomini, Pabst Pius II: Ausgewählte Texte aus seinen Schriften*, ed and trans Berthe Widmer, Latin and German (Basel 1960)

President's Science Advisory Committee. *The World Food Problem*, 3 vols (Washington 1967)

Pulleyblank, E. G. 'Chinese Historical Criticism: Liu Chih-chi and Ssu-ma Kuang', in W. G. Beasley and E. G. Pulleyblank (eds), *Historians of China and Japan* (London 1961), 134–66

Reynold, Gonzague de. *Le monde grec et sa pensée* (Fribourg en Suisse 1944)

Richthofen, F. von. *Baron Richthofen's Letters, 1870–1872*, 2nd ed (Shanghai 1903)

Richthofen, Ferdinand von. *China*, 1 (Berlin 1877)

Rose, Albert H. *A Geography of International Relations* (Dayton, Ohio 1965)

Rougemont, Denis de. *The Idea of Europe*, trans Norbert Guterman (New York 1966)

Rubin, V. A. 'Problemy vostochnoy despotii v rabotakh sovetskikh issledovateley', *Narody Azii i Afriki*, no 4 (1966), 95–104

Saushkin, Yu. G. 'The Geographical Environment of Human Society', *Soviet Geography: Review and Translation*, 4 (December 1963), 3–19

Schlegel, Friedrich. *Studien zur Geschichte und Politik*; 7.1 of *Kritische Friedrich-Schlegel-Ausgabe*, ed Ernst Behler (Munich 1966)

Schram, Stuart. *Mao Tse-tung* (Baltimore nd)

Schram, Stuart, and d'Encausse, Hélène Carrère. *Le Marxisme et l'Asie, 1853–1964* (Paris 1965)

Schulin, Ernst. *Die weltgeschichtliche Erfassung des Orients bei Hegel und Ranke*, Veröffentlichungen des Max-Planck-Instituts für Geschichte, 2 (Göttingen 1958)

Schurmann, Franz. 'Chinese Society', in David L. Sills (ed), *International Encyclopedia of the Social Sciences*, 2 (1968), 408–25

Schwind, Martin. 'Die geographischen "Grundlagen" der Geschichte bei Herder, Hegel und Toynbee', *Erdkunde*, 14 (March 1960), 3–10

Semple, Ellen Churchill. *Influences of Geographic Environment, On the Basis of Ratzel's System of Anthropo-Geography* (New York 1911)

Skinner, G. William. 'Marketing and Social Structure in Rural China', parts 1, 2 and 3, *Journal of Asian Studies*, 24 (November

1964), 3–43; (February 1965), 195–228; (May 1965), 363–99

Smith, Arthur H. *Chinese Characteristics*, 13th ed (New York 1894)

Snow, Edgar. 'Population Control in China: An Interview with Chou En-lai', in Larry K. Y. Ng (ed), *The Population Crisis* (Bloomington, Indiana 1965), 99–103

Sorokin, Pitirim. *Contemporary Sociological Theories* (1928, reprinted New York 1964)

Sprout, Harold and Margaret. *The Ecological Perspective on Human Affairs* (Princeton 1965)

Stalin, J. 'Dialectical and Historical Materialism' (September 1938), in idem, *Problems of Leninism* (Moscow 1945), 569–95

Strabo. *The Geography of Strabo*, with an English translation by Horace Leonard Jones (London and New York 1917–32)

Sun Ching-chih et al. *Severnyi Kitay; ekonomicheskaya geografiya* (Moscow 1958) (trans from Chinese by B. Sh. Zabirov et al)

Taeuber, Irene B. 'Asian Populations; the Critical Decades', presented to the Committee on the Judiciary, House of Representatives, 13 September 1962, in Larry K. Y. Ng (ed), *The Population Crisis* (Bloomington, Indiana 1965), 72–87

Taeuber, Irene B. 'China's Population: Riddle of the Past, Enigma of the Future', in Albert Feuerwerker (ed), *Modern China* (Englewood Cliffs 1964), 16–26

Tatham, George. 'Environmentalism and Possibilism', in Griffith Taylor (ed), *Geography in the Twentieth Century*, 3rd ed (New York and London 1957), 128–62

Taylor, Griffith (ed). *Geography in the Twentieth Century*, 3rd ed (New York and London 1957)

Teggart, Frederick J. *Rome and China; a Study of Correlations in Historical Events* (Berkeley, California 1939)

Thomas, Franklin. *The Environmental Basis of Society; A Study in the History of Sociological Theory* (London and New York 1925)

Thompson, Laurence G. *Ta T'ung Shu; The One-World Philosophy of K'ang Yu-wei* (London 1958)

Thompson, Warren S. *Population and Progress in the Far East* (Chicago 1959)

Thomson, J. Oliver. *History of Ancient Geography* (Cambridge 1948)

Timasheff, Nicholas S. *Sociological Theory: Its Nature and Growth*, 3rd ed (New York 1967)

Tozer, H. F. *A History of Ancient Geography*, 2nd ed (New York 1964)

Trewartha, Glenn T. 'Ratio Maps of China's Farms and Crops', *Geographical Review*, 28 (1938), 102–11

Turgot, Anne-Robert-Jacques. *Oeuvres de Turgot*, 2 vols (Paris 1844)

US Government, Department of State. *United States Relations with China*, Department of State Publication 3573, Far Eastern Series 30 (Washington 1949)

Varga, E. *Ocherki po problemam politekonomii kapitalizma* (Moscow 1965)

Wallach, Richard. *Das abendländische Gemeinschaftsbewusstsein im Mittelalter*, Beiträge zur Kulturgeschichte des Mittelalters und der Renaissance, ed Walter Goetz, Band 34 (Leipzig and Berlin 1928)

Wang Chun-heng. *A Simple Geography of China* (Peking 1958)

Wang, K. P. 'The Mineral Resource Base of Communist China', in Joint Economic Committee, Congress of the United States, *An Economic Profile of Mainland China*, 1 (Washington 1967), 167–95

Wang, K. P. 'Rich Mineral Resources Spur Communist China's Bid for Industrial Power', *Mineral Trade Notes*, Special Supplement no 59 (March 1960)

Watt, Kenneth E. F. *Ecology and Resource Management; A Quantitative Approach* (New York 1968)

Wetter, Gustav A. *Dialectical Materialism: A Historical and Systematic Survey of Philosophy in the Soviet Union*, trans Peter Heath (New York 1958)

Wetter, Gustav A. *Soviet Ideology Today*, trans Peter Heath (London 1966)

Whittlesey, Derwent. *Environmental Foundations of European History* (New York 1949)

Whittlesey, Derwent. 'The Regional Concept and the Regional Method', in Preston E. James and Clarence F. Jones (eds), *American Geography: Inventory and Prospect* (1954), 19–68

Wilhelm, Richard (original trans to German). *The I Ching, or Book of Changes*, trans into English by Cary F. Baynes, 3rd ed (Princeton 1967)

Wissmann, Hermann von. 'On the Role of Nature and Man in Changing the Face of the Dry Belt of Asia', in William L. Thomas, Jr (ed), *Man's Role in Changing the Face of the Earth* (Chicago 1956), 278–303

Wittfogel, Karl A. *Das erwachende China: ein Abriss der Geschichte und der gegenwärtigen Probleme Chinas* (Vienna 1926)

Wittfogel, Karl A. 'The Foundations and Stages of Chinese Economic History', *Zeitschrift für Sozialforschung*, 4 (1935), 26–60

Wittfogel, Karl August. 'Geopolitika, geograficheskiy materializm i marksizm', parts 1, 2 and 3, *Pod Znamenem Marksizma* (1929): nos 2–3 (February–March), 16–42; no 6 (June), 1–29; nos 7–8 (July–August), 1–28

Wittfogel, Karl A. 'Ideas and the Power Structure', in Wm Theodore de Bary and Ainslie T. Embree (eds), *Approaches to Asian Civilizations* (New York 1964), 86–97

Wittfogel, Karl A. 'Imperial China—a "Complex" Hydraulic (Oriental) Society', in John Meskill (ed), *The Pattern of Chinese History* (Boston 1965), 85–95

Wittfogel, Karl A. 'Die natürlichen Ursachen der Wirtschaftsgeschichte', *Archiv für Sozialwissenschaft und Sozialpolitik*, 67 (1932), 466–92, 579–609, 711–31

Wittfogel, Karl A. *New Light on Chinese Society; an Investigation of China's Socio-economic Structure* (New York 1938)

Wittfogel, Karl A. *Oriental Despotism; A Comparative Study of Total Power* (New Haven 1959)

Wittfogel, Karl A. 'Results and Problems of the Study of Oriental Despotism', *Journal of Asian Studies*, 28 (February 1969), 357–65

Wittfogel, Karl A. 'Die Theorie der orientalischen Gesellschaft', *Zeitschrift für Socialforschung*, 7 (1938), 90–122

Wittfogel, Karl A. *Wirtschaft und Gesellschaft Chinas; Versuch der wissenschaftlichen Analyse einer grossen asiatischen Agrargesellschaft*, 1 Teil: Produktivkräfte, Produktions- und Zirkulationsprozess, 3 (Leipzig 1931)

Wittfogel, Karl A., and Feng Chia-sheng. *History of Chinese Society: Liao (907–1125)*, Transactions of the American Philosophical Society, new series, 36 (1946) (Philadelphia 1949)

Wittkower, Rudolf. 'Marvels of the East: A Study in the History of Monsters', *Journal of the Warburg and Courtauld Institutes*, 5 (1942), 159–97

Wright, John Kirtland. *The Geographical Lore of the Time of the Crusades; A Study in the History of Medieval Science and Tradition in Western Europe*, American Geographical Research Series no 15 (New York 1925)

Wu T'ing-ch'iu. 'Establish a New System of World History', *Kuang ming jih pao* (7–10 April 1961)

Yang, C. K. *Religion in Chinese Society* (Berkeley California, 1961)

Yule, Henry. *Cathay and the Way Thither*, new ed revised by Henri Cordier, Hakluyt Society second series no 38 (1915) (Taipei 1966)

Zhukov, E. M. (E. Joukov). 'Des principes d'une histoire univer-
selle', *Cahiers d'Histoire Mondiale*, 3 (1956), 527–35
Zhukov, E. M. (ed). *Vsemirnaya Istoria v 10 tomakh*, 1 (Moscow 1955)
Zimmermann, Erich W. *Introduction to World Resources*, ed by Henry
L. Hunker (New York 1964)

CHINESE WORKS

Dynastic histories as in *Ch'in ting erh shih wu shih*, 1739, Ya chou
t'ung wen chü, 1879
SPPY: *Szu pu pei yao* (Shanghai 1920–35)
SPTK: *Szu pu ts'ung k'an* (Shanghai, republican period)
TS: *Ts'ung shu chi ch'eng* (Shanghai and Changsha 1935–9)

Chang Ch'i-yün 張其昀, *Chung kuo ti li hsüeh yen chiu* 中國地理學研究 [Studies on Chinese Geography], vol 1 (Taipei 1955)

Chang Han-ying 張含英, *Li tai chih Ho fang lüeh shu yao* 歷代治河方略述要 [Essentials of Yellow River Control Policies Through the Ages] (Chungking and Shanghai 1945 and 1946)

Cheng Chao-ching 鄭肇經, *Chung kuo shui li shih* 中國水利史 [History of Water Conservancy in China] (Taipei 1966)

Cheng Ch'iao 鄭樵, *T'ung chih lüeh* 通志略 [Appendices to the General Treatise], c 1150, SPPY vols 98–9

Chin Yü-fu 金毓黻, *Chung kuo shih hsüeh shih* 中

國史學史 [A History of Chinese Historiography], 2nd (Peking 1962)

Ch'iu Chün 丘濬, *Ta hsüeh yen i pu* 大學衍義補 [Supplement to the 'Expansion of the "Great Learning" '], part in T'u shu chi ch'eng 9 (Huang chi tien), 273 tsung lun, 8a–28a

Chou i 周易 [The Book of Changes] (first half of first millenium BC), SPTK vol I

Chu Chien-nung 朱劍农, *T'u ti fei li ching chi yüan li* 土地肥力經济原理 [Economic Principles of Soil Fertility] (Shanghai 1964)

Chung kuo kung ch'an tang ti liu tz'u ch'üan kuo ta hui i chüeh an
中國共產黨第六次全國大會
議決案 [Resolutions of the Sixth National Congress
of the Chinese Communist Party] (Moscow? 1928)

Li chi chu su 禮記注疏 [The Book of Rites]
(1st century BC and earlier), SPPY vols 22-4

Li Lin-ku 李林谷 'She hui sheng ch'an fang shih ho
jen k'ou wen t'i' 社会生产方式和人口
問題 [The Social Mode of Production and the Population
Problem], *Hsin chien she* 新建設, no 4 (1960), 49-55

Li Tsung-t'ung 李宗侗, Chung kuo shih hsüeh shih
中國史學史 [A History of Chinese Historiography],
2nd ed (Taipei 1962)

Liang Ch'i-ch'ao 梁啟超, *Chung kuo chin san pai nien hsüeh shu shih* 中國近三百年學術史 [A History of Chinese Scholarship in the Last Three Hundred Years] (1st ed 1935, reprinted Taipei 1958)

Liu Chih-chi 劉知幾, *Shih t'ung* 史通 [Complete History], 710, SPTK vols 163–4

Liu Chou 劉畫, *Liu tzu* 劉子 [Master Liu] (6th century), TS vol 595

Liu T-nien 刘大年, 'Ya chou li shih tzen yang p'ing chia?' 亞洲历史怎样評价 [How is Asian History to be Evaluated], *Li shih yen chiu* 历史研究, no 3 (1965), 1–24

Mao Tse-tung 毛澤東, *Mao Tse-tung hsüan chi* 毛澤東選集 [Selected Works of Mao Tse-tung] (Peking 1961 (vols 1, 2, and 3) and 1960 (vol 4))

Meng tzu 孟子 [Mencius] (*c* 290 BC), SPTK vol 23

Shang shu cheng i 尚書正義 [The Book of Documents] (pre-Han), SPPY vol 15

Shen Ping 申丙, *Huang ho t'ung k'ao* 黃河通考 [Treatise on the Yellow River] (Taipei 1960)

Shen Tsung-han 沈宗瀚, *Chung kuo nung yeh tzu yüan* 中國農業資源 [Agricultural Resources of China], 3 vols (Taipei 1952)

Shih Nien-hai 史念海, *Ho shan chi* 河山集 [Essays on the Historical Geography of China] (Peking 1963)

Sung Hsi-shang 宋希尚 et al, *Chung kuo ho ch'uan chih* 中國河川誌, [Rivers and Streams of China], 2 vols (Taipei 1955)

T'ien chin shih wen hua chü k'ao ku fa chüeh tui 天津市文化局考古发掘队 [Archaeological Excavation Brigade of the Tientsin Municipal Bureau of Culture], 'Po hai wan hsi an k'ao ku tiao ch'a ho hai an hsien pien ch'ien yen chiu' 渤海灣西岸考古調查和海岸線变迁研究 [Archaeological Investigation of the West Coast of the Po-hai and Study of Changes in the Coastline], *Li shih yen chiu* 历史研究 (1966), no 1, pp 52-62

Tsou Pao-chün 鄒豹君 *Ti hsüeh t'ung lun* 地學通論 [Geography], 8th ed (Taipei 1965)

Tu Yu 杜佑, *T'ung tien* 通典 [Comprehensive Institutes] (*c* 812) (Taipei 1959)

159

Tz'u hai 辭海 [Ocean of Words], 2 vols, 5th Taiwan ed (Taipei 1961)

Wang I-yai 王益厓, 'Chung kuo ti li hsüeh shih' 中國地理學史 [History of Chinese Geography), in Lin Chih-p'ing 林致平 et al, *Chung kuo k'o hsüeh shih lun chi* 中國科學史論集 [Collected Discussions on the History of Chinese Science], vol 1 (Taipei 1958), 67–121

Wang Yung 王庸, *Chung kuo ti li hsüeh shih* 中國地理學史 [A History of Chinese Geography] (Changsha 1938)

Wang Yung 王庸, *Chung kuo ti li t'u chi ts'ung k'ao* 中國地理圖籍叢考 [Collected Studies of Chinese Geographic Maps and Writings] (1947, reprinted Shanghai 1956)

ACKNOWLEDGEMENTS

Most of this book was written from 1966 to 1969 while I was at the East Asian Institute of Columbia University. I am grateful to the Institute and to my associates there for the various help and stimulation they gave me, including grants for travel to Taiwan and Hong Kong in the summer of 1966 and for writing in the summer of 1969. I am especially indebted to Morton Fried for his ideas and encouragement, and to Mervyn Adams Seldon, Dale Anderson Finlayson, and Herschel Webb for several years' friendly support on publication problems; more recently, Amy Heinrich has been helpful. Some of the themes were developed in connection with an Oriental Studies course at Columbia originally organised by Karl A. Wittfogel which I helped teach, first with Ainslee Embree and then with Leonard A. Gordon; the association with them and the students in the course was very useful, as was the chance to hear Professor Wittfogel who addressed the class on several occasions. George B. Cressey supervised my MA at Syracuse University, and though I criticise his outlook he was a valuable teacher. Basic geographic questions about Chinese society and culture were first pointed out for me by Rhoads Murphey at the University of Washington in the years 1959 to 1963, and it was under his guidance that I began studying much of what has gone into this book.

Andrew L. March

Denver

INDEX

STUDIES OF THE EAST
ASIAN INSTITUTE

Barnett, A. Doak. *Cadres, Bureaucracy, and Political Power in Communist China* (New York 1967)

Borg, Dorothy and Shumpei Okamoto (eds) (assisted by Dale K. A. Finlayson). *Pearl Harbor as History: Japanese-American Relations, 1931–1941* (New York 1973)

Chu, Samuel. *Reformer in Modern China: Chang Chien, 1853–1926* (New York 1965)

Clubb, O. Edmund. *China and Russia: 'The Great Game'* (New York 1971)

Curtis, Gerald L. *Election Campaigning Japanese Style* (New York 1971)

Dae-Sook Suh. *Documents of Korean Communism, 1918–1948* (Princeton 1970)

——. *The Korean Communist Movement, 1918–1948* (Princeton 1967)

Gurtov, Melvin. *The First Vietnam Crisis* (New York 1967)

Harrison, James P. Jr. *The Communists and Chinese Peasant Rebellions: A Study in the Rewriting of Chinese History* (New York 1969)

Hsiung, James C. *Law and Policy in China's Foreign Relations: A Study of Attitudes and Practice* (New York 1972)

Hu, C. T. (ed). *Aspects of Chinese Education* (New York 1970)

Huang Hsiao, Katherine. *Money and Monetary Policy in Communist China* (New York 1971)

Koji Taira. *Economic Development and the Labor Market in Japan* (New York 1970)

Koya Azumi. *Higher Education and Business Recruitment in Japan* (New York 1969)

March, Andrew L. *The Idea of China: Myth and Theory in Geographic Thought* (Newton Abbot 1974)

Morley, James William (ed). *Japan's Foreign Policy, 1868–1941: A Research Guide* (New York 1973)

Nakamura, James I. *Agricultural Production and Economic Development in Japan, 1873–1922* (Princeton 1966)

Passin, Herbert. *Japanese Education: A Bibliography of Materials in the English Language* (New York 1970)

——. *Society and Education in Japan* (New York 1965)

Ping-ti Ho. *The Ladder of Success in Imperial China* (New York 1962)

Ryan, Marleigh. *Japan's First Modern Novel: Ukigumo of Futabatei Shimei* (New York 1967)

Shuh-hsin Chou. *The Chinese Inflation, 1937–1949* (New York 1963)

Shumpei Okamoto. *The Japanese Oligarchy and the Russo-Japanese War* (New York 1970)

Steslicke, William E. *Doctors in Politics: The Political Life of the Japan Medical Association* (New York 1973)

Thayer, Nathaniel B. *How the Conservatives Rule Japan* (Princeton 1969)

Thurston, Donald Ray. *The Japan Teachers' Union: A Radical Interest Group in Japanese Politics* (Princeton 1973)

Titus, David Anson. *Palace and Politics in Prewar Japan* (New York 1973)

Varley, H. Paul. *Imperial Restoration in Medieval Japan* (New York 1971)

——. *Japanese Culture: A Short History* (New York 1973)

Watt, John R. *The District Magistrate in Late Imperial China* (New York 1972)

Webb, Herschel. *The Japanese Imperial Institution in the Tokugawa Period* (New York 1968)

Webb, Herschel (assisted by Marleigh Ryan). *Research in Japanese Sources: A Guide* (New York 1965)

Weinstein, Martin E. *Japan's Postwar Defense Policy, 1947–1968* (New York 1971)

167